# STOP OVER COMPLICATING PYTHON

programming doesnt have to be complicated

Veasna M. Mam

# Contents

# DISCLAIMER NOTICE

This book is a great starting point for learning Python, but it's important to keep in mind that real-world experience is key. Exercises are included with each chapter to help readers get started. Keep in mind that the programming field is constantly evolving, and while the information in this book is accurate as of its publication date, you should always stay up-to-date with the latest developments and best practices. Ultimately, how you use the knowledge gained from this book is up to you, and neither the author nor the publisher can be held responsible for any consequences resulting from your use of Python or the information provided in this book.

# ABOUT ME AS AN AUTHOR?

I am an alumnus of Fort Hays State University with a Bachelor of Science in Computer Science. I am also a certified ISTQB software tester, PECP Python certificate holder, and AWS-certified cloud practitioner. I have been programming for seven years, and Python is my preferred programming language. I have built several projects, including social media platforms, personal blogs, blockchain, and apps.

I discovered my passion for programming when I took an Introduction to Computer Science class in my first year of college. I was originally pursuing a degree in business, but I fell in love with programming and never looked back. Now, years later, I have decided to teach others the art of coding and help programmers become coding experts.

# INTRODUCTION TO CODING:
# UNDERSTANDING THE BASICS

I n today's digital world, coding has become a fundamental skill. Whether it's developing software, creating websites, building apps, or programming robots, coding is what makes technology work. Learning how to code opens up endless possibilities and helps you understand how the digital systems we interact with daily operate. But before diving into the complexities of coding, it's important to start with the basics—what coding is, why it matters, and some key terminology and concepts you'll encounter along the way.

## WHAT IS CODING?

Coding, sometimes called programming, is the process of using a programming language to give a computer instruction. These instructions tell the computer what tasks to perform and how to execute them. Computers, in their most basic form, are powerful machines capable of processing vast amounts of data and solving problems. However, they only understand very specific instructions written in a language they can process. This is where coding comes in.

Coding translates human ideas into a language the computer can understand. It's the act of writing these instructions in a specific format, known as code, to achieve a desired outcome. Every program or software you use—from a simple calculator app on your phone to complex systems like operating systems and video games—was built using code.

When starting to learn coding, it's important to familiarize yourself with some basic concepts and terminology that you'll frequently encounter. First is the idea of a program, which is essentially a set of instructions written in code that tells the computer how to perform a specific task. These programs can range from simple scripts that automate tasks to complex software applications.

Next, you'll encounter algorithms, which are sets of steps designed to solve a problem or complete a task. An algorithm is essentially the logic behind the program, detailing the sequence of actions the code will follow. For example, an algorithm for sorting a list of numbers would describe the method for comparing and arranging those numbers in order.

A key component of coding is the syntax of the programming language you're using. Syntax refers to the set of rules that define the structure of the code. Just as grammar rules dictate how sentences are constructed in natural languages, syntax dictates how code must be written for a program to function correctly. Even small mistakes in syntax, such as missing a semicolon or misplacing parentheses, can cause errors in the program.

You'll also need to understand the concept of variables, which store data that the program can manipulate. Variables can hold different types of data, such as numbers, text, or lists. For example, a variable in a program might store a user's age, a product's price, or a list of items in a shopping cart. Being able to use variables effectively is key to writing efficient and dynamic programs.

Functions are another important concept in coding. A function is a block of code that performs a specific task and can be reused throughout a program. This not only helps keep the code organized but also makes it more efficient by avoiding repetition. Functions can take input (parameters), perform operations, and return a result.

Lastly, coding also involves understanding loops and conditional statements. A loop allows a program to repeat a block of code multiple times, which is useful when you need to perform repetitive tasks. Conditional statements, on the other hand, allow programs to make decisions based on certain conditions. These "if-then" statements enable a program to choose different paths of action depending on the input or environment.

Learning to code opens doors to endless possibilities. It equips you with the ability to create solutions to problems, understand how digital systems work, and communicate with the technology that shapes our world. The basics of coding—understanding programs, algorithms, syntax, variables, functions, loops, and conditionals—are the building blocks upon which more complex coding skills are developed. With these foundations, you can start to explore the incredible potential of coding and the impact it can have on the world around you.

## WHY IS CODING IMPORTANT?

Coding is not just for professional software developers. As our lives become increasingly reliant on technology, understanding the basics of coding is becoming essential in many fields, including business, healthcare, education, and engineering. Here are a few reasons why coding is so important:

1. **Problem-solving Skills**: Coding teaches you to break down complex problems into smaller, manageable tasks. This analytical mindset is useful in nearly every aspect of life.

2. **Innovation**: With coding, you can create new products, services, and technologies, driving innovation across industries.

3. **Automation**: Coding allows us to automate tasks, making processes faster, more efficient, and often more accurate.

4. **Career Opportunities**: Proficiency in coding opens up a wide range of job opportunities in tech and non-tech fields alike, from software development to data science and artificial intelligence.

Now that we've covered what coding is and why it's important let's dive into some key terminology and concepts that are fundamental to understanding programming.

## KEY CODING TERMINOLOGY

Before starting to write code, it's important to familiarize yourself with some basic coding terminology. These are the building blocks of programming and will help you navigate the world of coding more easily.

1. **Algorithm**: An algorithm is a step-by-step procedure or formula for solving a problem. In coding, algorithms are used to create programs that can execute specific tasks. Think of it as a recipe: a set of instructions that the computer follows to achieve a goal.

2. **Programming Language**: A programming language is a formal language used to communicate with computers. There are many different programming languages, each with its own syntax and use cases. Some popular programming languages include:

- **Python**: Known for its simplicity and readability, Python is often recommended for beginners.

- **JavaScript**: Widely used for web development to create interactive websites.

- **Java**: A versatile language used for building everything from mobile apps to enterprise-level systems.

- **C++**: Known for its speed and efficiency, often used in game development and high-performance software.

3. **Syntax**: Syntax refers to the set of rules that define the structure of a programming language. Just like grammar in a spoken language, syntax dictates how code must be written for it to be understood by the computer. Every programming language has its own syntax; writing code that doesn't follow the correct syntax will result in errors.

4. **Variable**: A variable is a placeholder used to store data that can be referenced and manipulated in a program. Variables allow you to save information, such as numbers or text, and use it later in your code.

**For example:**

```
x = 5
name = "Alice"
```

In this case, x is a variable that stores the number 5, and the name is a variable that stores the string **"Alice."**

**Function**: A function is a block of reusable code designed to perform a specific task. Functions help you avoid repetition by allowing you to write a set of instructions once and reuse them whenever needed.

**For example:**

```
def greet():
print("Hello, World!")
greet()
```

In this example, the function **greet()** contains a simple print statement that displays **"Hello, World!"** when called.

**Loop**: Loops are used to repeat a block of code multiple times. They're essential for tasks that require repetition, such as processing each item in a list or counting down from a number. The two main types of loops are:

- **For Loop**: Used to iterate over a sequence (like a list or a range of numbers).

- **While Loop**: Repeats a block of code as long as a specified condition is true.

*Note: Please refer to chapter 6 for more details on for loops and while loops on page 79.*

**Conditional Statements**: Conditional statements allow programs to make decisions based on certain conditions. The most common conditional statement is the **if statement**, which executes a block of code if a condition is true.

**For example:**

```
age = 18
if age >= 18:
    print ("You are an adult.")
```

In this case, the program will print "You are an adult" if the value of age is 18 or higher.

**Data Types**: In coding, data types represent the different kinds of data that can be used and manipulated. Common data types include:

- **Integer**: Whole numbers (e.g., 1, 100, -5).

- **Float**: Decimal numbers (e.g., 3.14, 0.99).

- **String**: Text data enclosed in quotes (e.g., "Hello, World!").

- **Boolean**: Represents True or False values, often used in decision-making.

**Array (or List)**: An array or list is a collection of elements (e.g., numbers, strings) stored in a specific order. Lists are useful when you need to store multiple values in a single variable. For example:

```
fruits = ["apple", "banana", "cherry"]
```

1. **Object-Oriented Programming (OOP)**: OOP is a programming paradigm that uses objects and classes to structure code. In OOP, code is organized around objects, which are instances of classes. A class defines a blueprint for objects, and objects are individual instances of those classes with specific properties and behaviors.

## HOW DO COMPUTERS UNDERSTAND CODE?

Computers do not understand human languages like English or Spanish. Instead, they understand machine language, which is a series of binary codes made up of 0s and 1s. When you write code in a high-level programming language like Python or Java, that code needs to be translated into machine language so the computer can execute it. This translation happens through two main processes:

1. **Compilation**: In some programming languages like C++ or Java, code is compiled before execution. A compiler translates the entire code into machine language at once, and then the computer runs the compiled code.

2. **Interpretation**: In other programming languages like Python, the code is interpreted line by line as it is executed. An interpreter reads the code and translates it into machine language on the fly.

Both methods allow the computer to understand and execute the instructions provided by the programmer.

## A BRIEF HISTORY OF CODING

Coding has evolved tremendously since the early days of computers. The first programmable machines, like Charles Babbage's Analytical Engine in the 19th century, relied on physical punch cards to input instructions. However, modern coding as we know it began in the mid-20th century with the invention of early electronic computers.

- **1940s**: The first digital computers were created, and machine language (binary code) was used to write programs. These early programs were incredibly complex and tedious to write.

- **1950s-1960s**: High-level programming languages like FORTRAN and COBOL were developed to make coding more accessible to humans. These languages allowed programmers to write code using more natural language constructs.

- **1970s**: The C programming language was developed, which became the foundation for many modern languages. C introduced more efficient ways to control hardware, making it a favorite for systems programming.

- **1980s-1990s**: Object-oriented programming (OOP) gained popularity with the rise of languages like C++ and Java, allowing for more modular and reusable code.

- **2000s and beyond**: Today, coding languages are more powerful, flexible, and easier to learn than ever before. Languages like Python and JavaScript have made coding accessible to a broader audience, and the rise of open-source communities has accelerated innovation in programming.

## HOW TO START CODING

Now that you have a foundational understanding of coding, you may be wondering how to get started. The good news is that learning to code has never been more accessible. Here are some steps to guide you on your coding journey:

1. **Choose a Programming Language**: Start with a beginner-friendly language like Python, known for its simple syntax and versatility.

2. **Set Clear Goals**: Whether you want to build a website, create an app, or automate tasks, having a clear goal will keep you motivated as you learn.

3. **Practice, Practice, Practice**: Coding is a skill, and like any skill, the more you practice, the better you become. Start with simple exercises and gradually move to more complex projects.

4. **Learn from Others**: Join coding communities, participate in forums, and explore open-source projects. Collaborating with others will help you learn faster and stay motivated.

So, Coding is more than just a technical skill—it's a way of thinking. It teaches you to break down complex problems, think logically, and come up with creative solutions. Whether you're coding for fun, as a hobby, or as part of your career, the possibilities are endless. Understanding the basics of coding is the first step on an exciting journey that will allow you to harness the power of technology and shape the future. So, grab your computer, pick a language, and start coding!

# CHAPTER 1: INTRODUCTION TO PYTHON

Python is one of the most widely used and versatile programming languages in the world today. From powering web applications and automating tasks to supporting machine learning models and analyzing vast amounts of data, Python has established itself as a go-to language for beginners and experts alike. Whether you are just getting started in coding or you are a seasoned developer looking for efficiency, Python has a lot to offer. This chapter introduces Python, explores what it can do, discusses why it stands out from other programming languages and offers a brief history of its development.

**WHAT IS PYTHON?**

Python is a high-level, interpreted programming language known for its simplicity and readability. Developed in the late 1980s by Guido van Rossum, Python was designed to make programming easier and more accessible. Its clean syntax allows programmers to write code that is easy to understand, making it an ideal language for both beginners and experts.

One of Python's primary strengths is its versatility. Python can be used for various applications, from web development and automation to data analysis and artificial intelligence. Because of its broad utility, Python is often referred to as a "general-purpose" programming language.

Python code is executed line by line, which allows for quick testing and iteration. Unlike languages such as C or Java, which require compilation before execution, Python's interpreted nature means that you can write and run code without waiting for lengthy compile times. This feature contributes to Python's reputation as a highly productive language.

## WHAT CAN PYTHON DO?

Python's range of applications is vast, and it continues to grow as more industries adopt the language. Below are some of the most popular uses of Python:

1. **Web Development** Python is widely used in web development, thanks to frameworks like Django, Flask, and Pyramid. These frameworks allow developers to build robust, scalable websites with minimal effort. Django, for example, is a high-level framework that enables developers to build web applications quickly and efficiently by automating many common web development tasks such as database management, authentication, and URL routing.

2. **Data Science and Machine Learning** Python has become the de facto language for data science and machine learning. Libraries such as NumPy, Pandas, Matplotlib, and Scikit-learn provide powerful tools for data manipulation, statistical analysis, and machine learning. Python's simplicity allows data scientists to focus more on their algorithms and analysis rather than the complexities of the code itself. Additionally, TensorFlow and PyTorch are widely used libraries for building neural networks and deep learning models.

3. **Automation and Scripting** Python is also an excellent tool for automating repetitive tasks. It's simple syntax and built-in modules allow for quick scripting of tasks like file manipulation, web scraping, and even system administration. Automation using Python can save hours of manual work by allowing scripts to perform tasks such as renaming multiple files, extracting data from websites, or automating data entry.

4. **Game Development** Although not as commonly associated with game development as languages like C++ or C#, Python still plays a role in this industry. Frameworks like Pygame allow developers to create 2D games quickly and easily. Python's ability to handle logic and algorithms effectively makes it an attractive choice for building prototypes or indie games.

5. **Artificial Intelligence (AI)** AI and machine learning are rapidly advancing fields, and Python's dominance in these areas cannot be overstated. Libraries like

Keras, OpenAI, and TensorFlow make it easier for developers to build intelligent systems that can recognize patterns, predict outcomes, and even perform natural language processing (NLP). Python's accessibility allows researchers and developers to experiment and prototype AI models at a faster pace than with many other languages.

6. **Scientific Computing** Python is a popular choice in scientific computing and academic research. Its libraries, like SciPy, SymPy, and Jupyter Notebooks, make it easy for scientists and researchers to perform complex mathematical calculations, simulations, and visualizations. Python's ability to handle large datasets and integrate with other languages like C++ makes it valuable in scientific computing projects.

7. **Internet of Things (IoT)** Python's lightweight nature makes it suitable for use in IoT projects. Microcontrollers like Raspberry Pi often use Python to interact with sensors and other devices, allowing for the development of smart home systems, robotics, and other connected devices.

## BENEFITS OF PYTHON OVER OTHER PROGRAMMING LANGUAGES

There are several reasons why Python stands out among other programming languages, such as C++, Java, and JavaScript. Below are the key benefits of using Python:

1. **Simplicity and Readability** Python's syntax is designed to be easy to read and write. Its code looks almost like plain English, which means developers spend less time deciphering it. Python also enforces indentation as part of its syntax, which naturally promotes good coding practices. This simplicity makes Python a favorite among beginners and experienced developers who want to write clean and efficient code.

2. **Large Ecosystem of Libraries and Frameworks** One of Python's greatest strengths is its extensive ecosystem. Python's package manager, pip, provides easy access to a vast array of libraries that cover everything from web development and machine learning to scientific computing and game development. This vast ecosystem means that developers can find tools and libraries to help them with virtually any problem, reducing the time spent on writing boilerplate code.

3. **Cross-Platform Compatibility** Python is a cross-platform language, meaning that Python code can run on any operating system, such as Windows, macOS, or

Linux. This flexibility allows developers to write code on one machine and run it on another without making significant changes. This feature is particularly beneficial for development teams working across different platforms.

4. **Community Support** Python's popularity has led to a large and active community of developers who contribute to its continued growth. Whether you need help debugging your code, learning a new framework, or seeking advice on best practices, there is an abundance of resources available. This community-driven support makes it easier to learn and troubleshoot issues.

5. **High Demand in the Job Market** Python's versatility and widespread use make it one of the most in-demand programming languages in the job market. Companies in tech, finance, healthcare, and many other industries are constantly looking for Python developers, particularly in the fields of web development, data science, and machine learning. Learning Python can significantly boost your employability.

6. **Integration Capabilities** Python can easily be integrated with other languages like C, C++, and Java, allowing developers to combine Python's simplicity with the performance advantages of other languages. This ability to integrate with legacy systems or high-performance modules makes Python a versatile tool in a developer's toolkit.

7. **Quick Development Time** Python's clean syntax and ease of use lead to faster development times. Developers can quickly prototype and test ideas without worrying about complex code structures. This speed makes Python an excellent choice for startups and companies looking to bring products to market quickly.

## A BRIEF HISTORY OF PYTHON

Python's development began in the late 1980s when Guido van Rossum set out to create a new programming language that was simple yet powerful. Van Rossum was inspired by the ABC language, which focused on teaching non-programmers to code. However, he wanted to create a language that was more flexible and open to extension. Python was officially released in 1991 as Python 0.9.0, and it has since evolved into the robust language we know today.

In its early years, Python gained traction in the academic and scientific communities due to its simplicity and versatility. Researchers found Python to be a powerful tool for writing scripts, analyzing data, and automating processes. Python's ability to easily integrate with other languages and systems made it popular in environments that required rapid prototyping.

As the internet became more prevalent in the late 1990s and early 2000s, Python's role in web development grew. The release of frameworks like Django and Flask cemented Python's place as a dominant language for building websites and web applications. During this time, Python's popularity skyrocketed, and it became a preferred language for companies like Google, YouTube, and Instagram.

In the 2010s, Python saw another surge in popularity with the rise of data science, machine learning, and artificial intelligence. Python's libraries for numerical computing, data manipulation, and machine learning, along with its straightforward syntax, made it the go-to language for data scientists and AI researchers.

Python has continued to evolve, with the most recent version, Python 3.x, introducing several improvements over the original Python 2.x series. Python 3's enhancements include better support for Unicode, improved syntax for printing, and more efficient memory handling. While some legacy systems still use Python 2.x, Python 3.x is now the standard for new projects.

Now, we know that Python's journey from a small project in the late 1980s to one of the most influential programming languages in the world is a testament to its simplicity, power, and adaptability. Whether you are building a website, analyzing data, automating tasks, or creating machine learning models, Python is an ideal choice. Its ease of use, vast ecosystem, and supportive community make it a valuable tool for both beginners and seasoned developers.

As you dive deeper into Python in the coming chapters, you will discover why Python has earned its place at the forefront of modern programming. Whether you are learning Python for personal projects or professional development, understanding its history and capabilities will give you the foundation to excel in this versatile and powerful language.

# Chapter 2: Setting Up Python

To effectively understand Python programming through this book, you need a computer with Python installed.

**How to install Python on a Windows machine?**

To kickstart your Python programming journey the initial step is to install the Python interpreter, along with a library of tested code modules on your computer. You can easily download these for free from the Python website at https://python.org/downloads. If you're using Windows, you have the option to select either the 32-bit or 64-bit installers.

**Here's how you can do it:**

**1. Go to the Python** website by visiting https://www.python.org/.

**2. Head over to the Downloads** section located on the homepage by clicking on the "Downloads" tab.

**3. Choose your version of Python** from the options for download. Typically, it's recommended to go for the latest stable version.

4. **Select the installer** for your operating system by scrolling down and locating it on the page. Since you are using Windows, opt for either the 32-bit or 64-bit installer based on your system architecture.

5. Click on the provided link to download the installer.

The download will begin automatically.

6. **Run the installer. Once the download is finished, find the downloaded file (in your Downloads folder) and double-click** on it to run the installer.

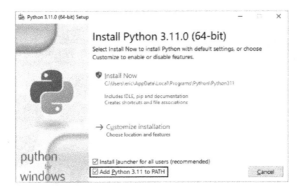

7. **Set up the installation**. In the installer, you'll see configuration options. Make sure to click the box that says, "Add Python to PATH". This ensures that Python is added to your system PATH, allowing you to run Python from the command line.

8. **Install Python**: Follow the instructions in the installer to complete the installation process. By default, Python will be installed in the C:\Python directory.

9. **Confirm the installation**; After installation, Command Prompt and type python *--version* to confirm that Python has been installed correctly. You should see the version number of the Python interpreter you installed.

**Please note**: If you have a 32-bit computer, use the 32-bit installer; otherwise, opt for the 64-bit installer if unsure about your computer's architecture.

**How to install Python on MacOS?**

*To install Python on MacOS I recommend installing it as software. Here are the steps to follow:*

As someone who has been using Mac for a time, I've observed that Python 2 is often pre-installed as a default program, with Unix support.

To check if MacOS is installed on your Apple-supported device, follow these steps:

**1. Open a terminal by navigating to Settings > Utilities > Terminal.**

**2.** Once the new terminal is up, type in the command:

*$ python3*

If you don't see the Python version displayed it means that Python is not installed on your MacOS system. To install Python, you can utilize homebrew by entering the following command in the terminal:

*$ brew install python3*

**Please note**: In case the initial method doesn't work, there's an approach to installing Python on MacOS.

*Follow these steps to download Python on macOS:*

**1. Visit the Python website** at https://www.python.org/.

**2. Click on the "Downloads" tab** located in the menu bar at the top of the homepage.

**3. Choose your desired version** of Python from the options; typically, it's recommended to select the stable version.

**4. Scroll down to find options** for different operating systems and select the macOS installer.

**5. Download the installer;** Click on the link to get the macOS installer. The download will start on its own.

**6. Run the installer.** Once the download finishes, find the downloaded file (in your Downloads folder) and double-click to run it.

**7. Set up the installation**. In the installer, you'll see setup options. Be sure to click on "Install for all users (recommended)". Add **Python 3.x** to PATH". These choices ensure that Python is installed and easily accessible from Terminal.

**8. Install Python:** Follow the instructions in the installer to finish installing Python. It will typically be installed in the /usr/local/bin directory by default.

**9. Confirm installation;** After installation, Terminal and type **python3** --version to confirm that Python was installed correctly. You should see the version number of your installed Python interpreter.

### How do I install Python on Linux?

Linux commonly comes with a version of Python (like 2.7) pre-installed. I suggest installing a version such as Python 3.x.

### *Here are the next steps to download Python on Linux:*

Getting Python on Linux is usually pretty simple, especially if you're using a package manager. Here's what you need to do:

**1. Open a terminal:** You can open a terminal on Linux distributions by searching for "Terminal" in the applications menu or by pressing **Ctrl + Alt + T.**

**2. Update your list of packages.** Use the command below to make sure your package lists are current:

*sql*

This command is specifically for Debian based distributions, like Ubuntu. For distributions, you might need to use a package manager or command.

**3. Install Python:** Use this command to install Python:

This will install Python 3, which is generally recommended because it includes updated versions. If you require Python 2, you can install it by replacing python3 with Python in the command.

**4.Check the installation.** Once the installation is done, confirm that Python has been installed correctly by running:

*python3 --version*

This command should show the version of the Python interpreter you installed.

**5. Optional: Add packages. Depending on your requirements,** you might want to install Python packages using the package manager.

For instance, you can set up pip, which is the Python package installer, by executing the following:

*sudo apt install python3-pip*

Also, swap python3 with Python if you have installed Python 2.

Begin utilizing Python: After installing Python, you can start using it by running the python3 command in the terminal. You can also. Execute Python scripts using a text editor or an integrated development environment (IDE).

# CHAPTER 3: INSTALLING THE PYTHON EXTENSION FOR VS CODE

O nce you've successfully installed Python on your computer, it's essential to choose an Integrated Development Environment (IDE) for a coding experience. While you can code Python in a text editor, like Notepad, it's recommended to opt for IDEs like PyCharm, Visual Studio Code (VSCode), Jupyter Notebook, or IDLE. Select an IDE that aligns with your preferences and needs. After picking an IDE, it's time to create your Python program. Open your chosen IDE. Start a Python file. A simple Python program involves printing **"Hello, World!"** on the screen.

**Use the code:**

```
print("Hello, World!")
```

Save this file with a `.py` extension like `hello.py` and run it from within your IDE to see "Hello, World!" displayed in the output. **Visual Studio Code**

VS Code is a high-quality text editor that's free of charge. It offers features supporting various programming languages and is beginner-friendly as well. Additionally, it can serve as a debugger.

As you get more comfortable using Python you can gradually transition to using VS Code for handling projects. VS Code is compatible with the following platforms: Windows 10 and 11 (64-bit) and macOS versions that receive Apple security updates. Typically, this includes the latest release and the two previous versions.

If you're eager to start coding, a good starting point would be to begin with VS Code. As your programming skills evolve, you may also want to explore editors or Integrated Development Environments (IDEs). In this section I'll walk you through the steps of setting up VS Code on your system. Ready to dive in?

**Note:** If you already have another text editor installed, feel free to use that.

**Setting Up VS Code**

Installing VS Code on Windows:

**1**. Download the setup file for VS Code from http;//code.visualstudio.com.

**2.** Click on the download for Windows option. Run the installation process as directed below.

**3.** To install VS Code, download the installer from https;//code.visualstudio.com. Once downloaded, locate it in your downloads folder. Move the Visual Studio Code installer into your applications folder. Finally, double-click on the installer file to initiate the installation process.

**Installing VS Code on Linux:**

You can easily install VS Code via the Ubuntu Software Center by clicking on the Ubuntu software icon in your menu bar.

When searching for Vscode, locate the Visual Studio Code App. Select Install. Once it's installed, find the VS code on your system. Open it.

**Installing the Python Extension for VS Code**

To begin coding in Python and make the most of its capabilities as a programming language, you'll need to add the Python extension. This extension will assist in writing, editing, and running Python scripts.

**To install the Python extension**, click on the manager icon located at the bottom left corner of VS Code within the application menu. In the menu that appears, Extensions.

Type "Python" in the search box. Choose the Python extension (make sure to pick Microsoft's version if multiple options are available). Click Install. Add any tools required for installation. If you receive a prompt indicating that you need to install Python even though it's already installed, you can disregard this message.

**Follow these steps to create and run your Python program:**

If you're using macOS, a pop-up may appear, prompting you to install command line developer tools. Click **"Install"** to proceed.

Before you kick off with coding make sure to set up a folder called **"python_work"** on your desktop or any preferred location. It's best to stick with lowercase letters and underscores of spaces, for your file and folder names to align with Python's naming conventions. Once that's done, fire up Visual Studio Code. Close the "Get Started" tab if it's still hanging around. Next create a file by clicking on "File" New File" or simply hit **"Ctrl N"** on Windows or "Command N" on macOS. Save this file as "hello_world.py" in your **"python_work"** directory. The ".**py**" extension signals to Visual Studio Code that this is a Python file, allowing it to highlight the code properly.

**After saving your file, enter the following line of code into the editor:**

```
print("Hello, world!")
```

To execute the program, click on **"Run"** in the menu or use shortcuts like **"F5"** for Windows or "fn F5" for macOS. You should see the output, which will display **"Hello, world!"** in the terminal. Well, done! You've just. Executed your first Python program!

# CHAPTER 4: PYTHON FOUNDATIONS

In Python programming, programmers must ensure that the input is provided directly by the user and that output is provided based on the inputs to have dynamic applications; the Python interpreter and all functions in your program can access the user input values.

In this chapter, I will show you a few examples and detailed programs to help you grasp and understand how to improve the user experience of the software you have created based on the input and output operations.

**The Importance of input values.**

Every application, be it web-based or a metaverse application, is heavily influenced by user input values. For instance, when you log in to Facebook, you are required to provide your email and password. These inputs are more than just strings of characters; they are the key to authenticating your account, ensuring that only you can access it. Facial recognition technology makes use of face data points as input to operate advanced applications. In today's world, every real-world application relies on user input data to provide a better user experience. For instance, let's say you have developed a Python application that is only intended for a mature audience, i.e., people above 18 years of age. In this case, you can utilize conditional input verification by requesting the user to enter their age. If the user is over 18, they will be granted access to the application. However, if the user's age is below 18, the application will remain inaccessible. Python evaluates all of the supported inputs to determine whether someone can access your software or not.

**Input Values in Python**

Python's **input()** function is a fundamental tool for creating interactive programs that take user input. It allows your program to pause and wait for the user to type in some data, which is then stored as a string in a variable.

**Example:**

```
name = input("What is your name? ")
print(f"Hello, {name}!")
```

**In this code:**

1. **name =input("What is your name?"):** The **input()** function displays the prompt "What is your name? " And waits for the user to enter their name. Whatever the user types are stored as a string in the variable **name**.

2. **print(f"Hello, {name}!"):** The **f-string** (formatted string literal) is used to create a personalized greeting message that incorporates the user's name retrieved from the **name** variable.

*When running the programming codes, it would look something like this.*

```
What is your name? John
Hello, John!
```

**Why Input Values Are Important Input values are essential in Python because they:**

- **Make programs interactive:** They allow users to participate in the program's execution by providing data, making it more engaging and user-friendly.

- **Enhance program functionality:** By incorporating user input, you can create programs that adapt to different situations or perform calculations based on the user's needs.

- **Practice basic data types:** Input values often involve working with strings, integers, or floating-point numbers, which are fundamental data types in Python. You'll learn how to handle different types of data and convert them as necessary.

- **Develop problem-solving skills:** When using input values, you need to think about how the user's input might affect your program's logic and how to handle potential errors (e.g., the user entering the wrong data type). This strengthens your problem-solving abilities.

- **Create a foundation for more advanced concepts:** Input values are a steppingstone to more complex topics like working with files, user interfaces, and network programming.

**Additional Tips:**

- **Prompt with clear instructions:** Guide the user with informative prompts that explain what kind of input is expected.

- **Consider data types:** If the user needs to enter a number, you might want to convert the input string to an integer or float using functions like **int()** or **float().**

- **Handle errors:** Anticipate potential errors (e.g., the user entering non-numeric characters when a number is expected). Use try-except blocks to manage these situations gracefully.

*By mastering input values, you'll gain a solid foundation for building interactive and adaptable Python programs.*

**What are User Prompts?**

A user prompt in Python is a message or question displayed to the user in order to request input from them while the program is running. It's a way for the program to communicate with the user and ask for specific information that the program needs to proceed or perform a task.

The **input()** function is commonly used to create user prompts in Python. When you use **input()**, you can pass a string argument that serves as the prompt message, asking the user for input. The program then waits for the user to type something, and press **Enter**. Whatever the user enters is typically stored in a variable for further processing by the program.

**For example:**

```
name = input("Please enter your name: ")
```

In this line of code, **"Please enter your name: "** is the prompt message. When the program runs, it will display this message to the user, and the user can type their name in response. Whatever they type will be stored in the variable **name.**

User prompts are commonly used in various types of applications, such as command-line interfaces, interactive programs, and forms where user input is needed. They help make programs more interactive and user-friendly by allowing users to provide input while the program is running.

### How to Write User Prompts?

In Python, you can use the **input()** function to prompt the user for input. Here's a simple example:

```
# Prompt the user for their name
name = input("Please enter your name: ")
# Print a greeting using the provided name
print("Hello, " + name + "! Nice to meet you.")
```

When you run this code, it will display the message "**Please enter your name**: " and wait for the user to type something. Whatever the user types will be stored in the variable **name**, and then the program will print a greeting using the provided name.

***Please note:***

The **input()** function is a versatile tool in Python, which is not only used to prompt the user with a single line of string but also to display multiple lines of strings, making it a powerful function in your Python toolkit. The **print()** function is a practical and straightforward way to display text on the screen. It is the go-to method for printing to a computer screen, giving you the confidence to communicate with your program. When you pass any input to the **print()** function, it is automatically converted to a string literal and displayed on the screen. While you do not need to know all of the **print()** function's arguments, understanding some of them can help you format your code and make it more visually appealing.

## What are string Literals?

A string literal is just a way to represent a string value directly in your code. It's like putting words or characters inside quotation marks. For example, "hello" and 'Python is fun' are both string literals.

They allow you to work with text in your Python programs. You can use them to store messages, words, sentences, or any other kind of textual data. So, whenever you see text enclosed in either single quotes (' ') or double quotes (" "), that's a string literal in Python.

In Python, a string literal is just a sequence of characters enclosed within either single quotes (' ') or double quotes (" ").

## Here's a simple example:

```python
# Single quotes
single_quoted_string = 'Hello, world!'
print(single_quoted_string)
# Double quotes
double_quoted_string = "Hello, world!"
print(double_quoted_string)
```

In this code, **single_quoted_string** and **double_quoted_string** are both string literals. They hold the text **"Hello, world!"** and are enclosed within single and double quotes, respectively.

Python doesn't care whether you use single or double quotes for string literals; you can use either as long as you're consistent within a string. So, these two lines of code are equivalent:

```python
single_quoted_string = 'Hello, world!'
double_quoted_string = "Hello, world!"
```

## Both create a string containing:

```python
"Hello, world!"
```

String literals can contain any characters, including letters, numbers, symbols, and even spaces. They're used to represent text data in Python programs.

*Note:* When working with data in code, it's important to ensure that the output is properly formatted for readability and ease of use. One way to achieve this is through the use of escape sequences. In particular, there are three popular escape sequences that can help you format output data in a new tab with no whitespaces or separators: **\t, \b,** and **\d**. The **\t** escape sequence is used to insert a tab character into the output, which can be useful for aligning columns of data. The **\b** escape sequence, on the other hand, is used to insert a backspace character, which can be used to delete the previous character in the output. Finally, the **\d** escape sequence is used to insert a digit into the output, which can be useful for generating numbered lists or other sequences of numbers. By using these escape sequences in your code, you can ensure that your output data is properly formatted and easy to read, making it more accessible and useful for yourself and others who may be working with the data.

**How to use an End Statement?**

An "**end statement**" is not a specific construct. However, I suspect you might be referring to the **end** parameter in the **print()** function in Python.

In Python, the **print()** function by default adds a newline character (**\n**) at the end of each print statement, which means each call to **print()** starts on a new line.

However, you can change this behavior using the **end** parameter.

Here's how the **print()**function with the **end** parameter works:

```
# Default behavior (prints on separate lines)
print("Hello")
print("World")
# Using the end parameter to stay on the same line
print("Hello", end=" ")
print("World")
# Using a different end character
print("Python", end=" - ")
print("Programming")
# Using an empty string as the end parameter
print("This is", end="")
print(" a single line output.")
```

The **end** parameter specifies what should be printed at the end of the output. By default, it's set to **'\n'**, meaning a newline character. You can change it to any string you want or even an empty string if you don't want anything to be printed at the end.

Here's a simple example demonstrating how to use the end parameter:

```python
# Without specifying end parameter
print("Hello", end=", ") # The end parameter is specified as ", "
print("world")
# Output: Hello, world (on the same line)
# Specifying end parameter
print("Hello", end=", ")
print("world")
# Output: Hello, world (on the same line)
```

In the first **print()** statement, the end parameter is explicitly set to **", "**, which prevent a newline and ensure the second **print()** statement continues on the same line.

In second **print()** statement, the end parameter is set to **", "**, which means that instead of moving to a new line after printing **"Hello"**, a **comma and a space** are appended. This allows the second **print()** statement to continue on the same line, printing "world" immediately after the comma and space.

**Numerical Values as input**

You can take numerical values as input from the user using the **input()** function and then convert the input to the desired numerical type, such as **int** (integer) or **float** (floating-point number). Here's how you can do it:

```python
# Taking integer input from the user
num_int = int(input("Enter an integer number: "))
# Taking floating-point input from the user
num_float = float(input("Enter a floating-point number: "))
# Displaying the input values
print("Integer number entered:", num_int)
print("Floating-point number entered:", num_float)
```

**In this example:**

- **input** ("Enter an integer number: ") prompts the user to enter an integer number.

- **int ()** converts the input value to an integer.

- Similarly, **input**("Enter a floating-point number: ") prompts the user to enter a floating-point number.

- **float()** converts the input value to a floating-point number.

**Here's how it works when you run the code:**

Enter an integer number: 42

Enter a floating-point number: 3.14

Integer number entered: 42

Floating-point number entered: 3.14

Keep in mind that **input()** always returns a string, so you need to convert it to the appropriate numerical type using **int()** or **float()** if you want to perform numerical operations with the input values.

**Example of an int:**

In Python, **int** is short for "**integer**," which represents whole numbers without any decimal point. Integers can be positive, negative, or zero. Here's a simple explanation with an example:

```
# Example of integers in Python
num1 = 10      # positive integer
num2 = -5      # negative integer
num3 = 0       # zero

# Displaying the integers
print("num1:", num1)
print("num2:", num2)
print("num3:", num3)
```

**In this example:**

- **num1** is assigned the value **10**, which is a positive integer.

- **num2** is assigned the value **-5**, which is a negative integer.

- **num3** is assigned the value **0**, which is zero.

You can perform various operations on integers, such as addition, subtraction, multiplication, division, and more. For example:

```python
# Arithmetic operations with integers
num1 = 10   # Example integer
num2 = -5   # Example integer
# Performing arithmetic operations
result_addition = num1 + num2        # Addition
result_subtraction = num1 - num2     # Subtraction
result_multiplication = num1 * num2  # Multiplication
result_division = num1 / num2        # Division
# Displaying the results
print("Addition:", result_addition)
print("Subtraction:", result_subtraction)
print("Multiplication:", result_multiplication)
print("Division:", result_division)
```

**When you run this code, you'll see the results of the arithmetic operations:**

Addition: 5

Subtraction: 15

Multiplication: -50

Division: -2.0

Integers are essential data types in Python and are widely used in various programming tasks. They're particularly useful when you're working with counting, indexing, or any situation where you need whole numbers.

**Example of a float and what is a Float?**

In Python, **float** is short for **"floating-point number."** **Floating-point numbers** represent real numbers with a decimal point. They can be positive or negative and can include fractional parts. Here's a simple explanation with an example:

```python
# Example of floats in Python
num1 = 3.14    # positive float
num2 = -0.5    # negative float
num3 = 2.0     # float with integer value
num4 = 1.25e-3 # scientific notation (float)
# Displaying the floats
print("num1:", num1)
print("num2:", num2)
print("num3:", num3)
print("num4:", num4)
```

**In this example:**

- **num1** is assigned the value **3.14**, which is a positive floating-point number.

- **num2** is assigned the value **-0.5**, which is a negative floating-point number.

- **num3** is assigned the value **2.0**, which is also a floating-point number but with an integer value.

- **num4** is assigned the value **1.25e-3**, which is in scientific notation representing **0.00125**.

You can perform various operations on floating-point numbers, similar to integers, such as addition, subtraction, multiplication, division, and more. For example:

```python
# Arithmetic operations with floats
num1 = 3.14 # Example float
num2 = -0.5 # Example float
# Performing arithmetic operations
result_addition = num1 + num2        # Addition
result_subtraction = num1 - num2     # Subtraction
result_multiplication = num1 * num2  # Multiplication
result_division = num1 / num2        # Division
# Displaying the results
```

```
print("Addition:", result_addition)
print("Subtraction:", result_subtraction)
print("Multiplication:", result_multiplication)
print("Division:", result_division)
```

When you run this code, you'll see the results of the arithmetic operations with floating-point numbers. Floating-point numbers are essential for representing a wide range of values, including those that require precision beyond whole numbers.

**Output:**

Addition: 2.64

Subtraction: 3.64

Multiplication: -1.57

Division: -6.28

**What is a Comment and how to use it?**

In Python, a comment is a piece of text within your code that is ignored by the Python interpreter. Comments are used to add explanatory notes or documentation to your code, which can help you and others understand the purpose of different parts of the code. They are not executed when the program runs.

Here's how you can create comments in Python:

*Example:*

```
# This is a single-line comment in Python
# Single-line comments start with the hash symbol (#) and continue until the end of the
line.
# Example:
print("Hello, world!")  # This is a comment explaining the print statement
"""
This is a multi-line comment in Python.
Multi-line comments are enclosed within triple quotes (''' or """).
They can span across multiple lines.
"""
# Example:
```

```
"""
This program calculates the sum of two numbers.
"""
# Defining two numbers
num1 = 10
num2 = 20
# Calculating the sum
total = num1 + num2
# Displaying the sum
print("The sum is:", total)
```

**In this example:**

- Single-line comments start with the # symbol and continue until the end of the line. They can be placed at the end of a line of code or on a line by themselves.

- Multi-line comments are enclosed within triple quotes (""" or """) and can span multiple lines. They are often used for longer explanations or documentation.

Comments are helpful for making your code more understandable and maintainable, especially when you or others revisit the code later. They can explain the purpose of variables, functions, or sections of code, making it easier to understand and debug.

**Reserved Keywords**

Reserved keywords in Python are predefined words that are part of the Python language syntax and have special meanings. These keywords cannot be used as identifiers (such as variable names, function names, or class names) because they are reserved for specific purposes within the language. Here's a simple explanation with some examples:

```
# Example of reserved keywords in Python
import keyword  # Importing the keyword module to access Python's reserved keywords
# Get the list of all reserved keywords in Python
reserved_keywords = keyword.kwlist   # The kwlist attribute contains all reserved keywords
print("List of reserved keywords in Python:")  # Print a header
print(reserved_keywords) # Print the list of reserved keywords
```

When you run this code, you'll get a list of all reserved keywords in Python:

**List of reserved keywords in Python:**

```
['False', 'None', 'True', 'and', 'as', 'assert', 'async', 'await', 'break', 'class', 'continue',
'def', 'del', 'elif', 'else', 'except', 'finally', 'for', 'from', 'global', 'if', 'import', 'in', 'is',
'lambda', 'nonlocal', 'not', 'or', 'pass', 'raise', 'return', 'try', 'while', 'with', 'yield']
```

These keywords are reserved because they have specific meanings within Python and are used to define the structure and logic of Python code. **For example:**

- **if, else, elif**: Used for conditional statements.

- **for, while**: Used for looping.

- **def, class**: Used for defining functions and classes, respectively.

- **import, from, as**: Used for importing modules and aliases.

- **True, False, None**: Represent Boolean values and absence of value, respectively.

If you try to use any of these reserved keywords as an identifier in your code, Python will raise a syntax error. So, it's important to avoid using reserved keywords as variable names or function names to prevent conflicts and ensure your code runs smoothly.

**Understanding Operators:**

In Python, operators are symbols or special characters that perform operations on operands (values or variables). They are used to manipulate data and perform various computations. Here are some common types of operators in Python, along with visual graphs and examples:

1. **Arithmetic Operators:**

- Addition (+)

- Subtraction (-)

- Multiplication (*)

- Division (**/**)

- Modulus (**%**)

- Exponentiation (**\*\***)

**This is a visual graph table of arithmetic operators along with an example below.**

| Operators | Meaning | Example | Result |
|-----------|---------|---------|--------|
| + | Addition | 4 + 2 | 6 |
| − | Subtraction | 4 − 2 | 2 |
| * | Multiplication | 4 * 2 | 8 |
| / | Division | 4 / 2 | 2 |
| % | Modulus operator to get remainder in integer division | 5 % 2 | 1 |
| ** | Exponent | 5**2 = $5^2$ | 25 |

```
# Example of arithmetic operators
num1 = 10
num2 = 3
addition = num1 + num2
subtraction = num1 - num2
multiplication = num1 * num2
division = num1 / num2
modulus = num1 % num2
exponentiation = num1 ** num2
print("Addition:", addition)
print("Subtraction:", subtraction)
print("Multiplication:", multiplication)
print("Division:", division)
print("Modulus:", modulus)
print("Exponentiation:", exponentiation)
```

2. **Comparison Operators:**

- Equal to (==)

- Not equal to (!=)

- Greater than (>)

- Less than (<)

- Greater than or equal to (>=)

- Less than or equal to (<=)

| OPERATORS | MEANING | EXAMPLE |
|-----------|---------|---------|
| == | EQUAL TO | 5 = =5 |
| != | NOT EQUAL TO | 26! = 3 |
| > | GREATER THAN | 100 > 67 |
| < | LESS THAN | 89 < 216 |
| >= | GREATER THAN OR EQUAL TO | 90 >= 54 |
| <= | LESS THAN OR EQUAL TO | 23 <=77 |

```
x = 10
y = 5
equal_to = x == y
```

```
not_equal_to = x != y
greater_than = x > y
less_than = x < y
greater_than_or_equal_to = x >= y
less_than_or_equal_to = x <= y
print("Equal to:", equal_to)
print("Not equal to:", not_equal_to)
print("Greater than:", greater_than)
print("Less than:", less_than)
print("Greater than or equal to:", greater_than_or_equal_to)
print("Less than or equal to:", less_than_or_equal_to)
```

3. Logical Operators:

- AND **(and)**

- OR **(or)**

- NOT **(not)**

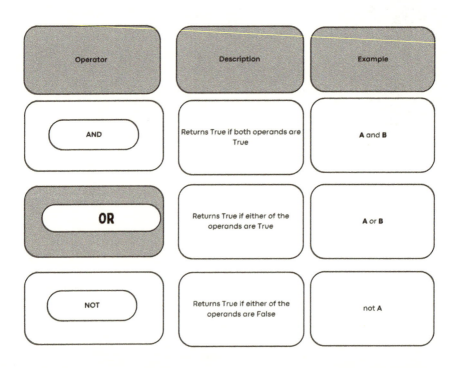

| Operator | Description | Example |
|----------|-------------|---------|
| AND | Returns True if both operands are True | A and B |
| OR | Returns True if either of the operands are True | A or B |
| NOT | Returns True if either of the operands are False | not A |

```
# Example of logical operators
a = True
b = False
logical_and = a and b
logical_or = a or b
logical_not = not a
print("Logical AND:", logical_and)
print("Logical OR:", logical_or)
print("Logical NOT:", logical_not)
```

4. Assignment Operators:

- Assignment (=)

- Addition assignment (+=)

- Subtraction assignment (-=)

- Multiplication assignment (*=)

- Division assignment (/=)

- Modulus assignment (%=)

- Exponentiation assignment (**=)

| OPERATORS | EXAMPLE | SAME AS |
|:---:|:---:|:---:|
| = | x = 5 | x = 5 |
| += | x+= 3 | x = x + 3 |
| -= | x-= 3 | x = x - 3 |
| *= | x*= 3 | x = x * 3 |
| /= | x/= 3 | x = x / 3 |
| %= | x%=3 | x = x % 3 |
| **= | x**= 3 | x = x **3 |

```
# Example of assignment operators
x = 10
x += 5   # Equivalent to: x = x + 5
print("Addition assignment:", x)
x -= 3   # Equivalent to: x = x - 3
print("Subtraction assignment:", x)
x *= 2   # Equivalent to: x = x * 2
print("Multiplication assignment:", x)
x /= 4   # Equivalent to: x = x / 4
print("Division assignment:", x)
x %= 3   # Equivalent to: x = x % 3
print("Modulus assignment:", x)
x **= 2   # Equivalent to: x = x ** 2
```

```
print("Exponentiation assignment:", x)
```

These are some of the commonly used types of operators in Python. They allow you to perform a wide range of operations on different types of data.

**What are Bitwise Operators?**

Bitwise operators in Python are used to perform operations at the bit level on binary representations of integers. They manipulate individual bits of integers directly, which can be useful in various scenarios such as low-level programming, cryptography, and optimization. Here are the different types of bitwise operators in Python, along with examples:

1. **Bitwise AND (&):**

- Sets each bit to 1 if both bits are 1.

```
# Example of bitwise AND operator
num1 = 5 # Binary: 101
num2 = 3 # Binary: 011
result = num1 & num2
print("Bitwise AND:", result) # Binary: 001 (Decimal: 1)
```

2. **Bitwise OR ( | ):**

- Sets each bit to 1 if at least one of the corresponding bits is 1.

```
# Example of bitwise OR operator
num1 = 5 # Binary: 101
num2 = 3 # Binary: 011
result = num1 | num2
print("Bitwise OR:", result) # Binary: 111 (Decimal: 7)
```

1. **Bitwise XOR ( ^ ):**

- Sets each bit to 1 if only one of the corresponding bits is 1.

```
# Example of bitwise XOR operator
```

```
num1 = 5 # Binary: 101
num2 = 3 # Binary: 011
result = num1 ^ num2
print("Bitwise XOR:", result) # Binary: 110 (Decimal: 6)
```

1. **Bitwise NOT ( ~ ):**

- Flips the bits, i.e., changes **1** to **0** and **0** to **1**.

```
# Example of bitwise NOT operator
num = 5  # Binary: 101
result = ~num print("Bitwise NOT:" , result)  # Binary: -110 (Decimal: -6)
```

1. **Left Shift (<<):**

- Shifts the bits to the left by a specified number of positions, filling the empty positions with zeros.

```
# Example of left shift operator
num = 5  # Binary: 101
result = num << 2
print("Left Shift:", result) # Binary: 10100 (Decimal: 20)
```

1. **Right Shift ( >> ):**

- Shifts the bits to the right by a specified number of positions, filling the empty positions with zeros or the sign bit.

```
# Example of right shift operator
num = 10  # Binary: 1010
result = num >> 1
print("Right Shift:", result)  # Binary: 101 (Decimal: 5)
```

**Example of exponentiation assignment operator:**

```
x = 2
```

```
x **= 3    # Equivalent to: x = x ** 3
print("Exponentiation Assignment:", x)   # Output: 8
```

These augmented assignment operators provide a convenient way to perform common operations and update variables in Python. They are concise and often result in cleaner code compared to writing out the full operation followed by an assignment statement.

*** 

**Exercises**

> 1. Write a program that prompts the user for their name and age, then prints out a message with their name and the year they will turn 100 years old.

2. Write a program that calculates the area of a circle given its radius.

3. Write a Python program that calculates the quotient and remainder of two given numbers without using the division (/) or modulo (%) operators.

4. Create a program that swaps the values of two variables without using a temporary variable.

5. Write a Python program to perform bitwise AND, OR, and XOR operations on two given integers.

6. Write a Python program that calculates the exponentiation of a given number.

# Chapter 5: Data Types and Variables

Data types play a role in Python as they define the characteristics of the data used in a program. Each data type has features and functions that cater to needs. Now let's dive into the data types in Python:

**1. Integers (int)** represent numbers without decimal points in Python. One notable aspect of integers is their precision allowing us to perform operations without concerns about overflow errors. They are commonly utilized in programming for tasks like counting, indexing, and arithmetic calculations. When we assign a value to an integer in Python, it creates an "in" object with that value, which can be modified if necessary. Integers are vital in programming as they enable reusing values and running loops to instruct computers on how many times to repeat tasks, simplifying processes and enhancing program efficiency.

**Let's see this code snippet at work:**

```
x = 2
x **= 3    # Equivalent to: x = x ** 3
print("Exponentiation Assignment:", x)   # Output: 8
```

In this snippet of code, we define two variables, **'x'** and **'y'**, set their values to **11** and **-84**, respectively, and showcase their values using the **'print()'** function.

**When we execute this code, the following results will be displayed:**

```
11
-84
```

Python is capable of handling numbers with up to ten digits without encountering any problems. Nevertheless, it's crucial to exercise caution when working with numbers to prevent any slowdown in the application. In situations, issues are unlikely to arise.

**2. Floating point numbers**, known as **floats**, represent numbers with points. However, due to the representation of values, floats are approximations that can introduce precision challenges in specific calculations. Despite this limitation, they find application in computations, financial modeling, and other tasks that demand decimal precision. Floats are instrumental in evaluating whether a comparison, between values yields False. Now, let's explore a program snippet to grasp how Booleans operates in Python:

**Program Code:**

```
a = 17
b = 45
print(a > b)
```

**When we run this code, it gives us the result:**

```
False
```

The value of 'a' (17) is not greater than the value of 'b' (45), so the Boolean expression "a > b" is determined as 'False'. Booleans come in handy when we are dealing with operations in our programs, helping us make choices based on conditions.

**Please note** Booleans play a role in programming as they assist us in making decisions based on certain circumstances. It's crucial to assess whether a condition holds true or false before utilizing it, which simplifies determining the steps for our program.

**3. Strings (str)**: Strings are sequences of characters enclosed within single (') or double (") quotations. They serve as data types used for storing information like names, messages and program codes. Python offers a range of string manipulation techniques for tasks such as combining strings, splitting them, formatting them, and more.

Understanding strings is paramount for developers to craft software. Representing data in strings is essential since computers comprehend information. Hence employing ASCII and Unicode encoding mechanisms holds importance.

**Let's see how this code works in action:**

```python
Color = "Blue"
print(Color)
```

**Output:**

```
Blue
```

Here, we have a text enclosed in quotes, which is the word "blue." This serves as an illustration of a string data type with the color" used to define it. The amount of memory occupied by a variable with a string data type is typically determined by its memory location and size. The bit count associated with a string data type is directly related to the number of characters it contains.

You can also utilize quotes when representing strings in your code, for **instance:**

```python
twitter_handle = '@POTUS'
print(twitter_handle)
```

**Output:**

```
@POTUS
```

**Examples showcasing strings:**

```
x = "Hello"
print(x)
```

**Output:**

```
Hello
```

```
# Employing quotes to denote the
string = '''This is a string defined using triple quotes allowing for multiple lines
content.'''
print(string)
```

**Output:**

```
This is a string defined using quotes, allowing for multiline content.
```

**Access Characters in Strings**

Here is an example that demonstrates ways to define strings in Python. Strings can contain characters, symbols, and new lines. Python also supports escape sequences like "\n" for creating line breaks. Additionally, Python provides built-in functions to manipulate characters within a string. Manipulating Characters in Strings in Python, the first character of a string is indexed as 0. Negative indexing can be used to access the character, while slicing allows you to extract parts of the string based on index positions. Understanding the indexing system starting from 0 is essential for accessing characters or specific sections of a string. To retrieve a character based on its position, you can follow these steps:

**Let's see how this code works in practice:**

```
# Define a string
```

```python
my_string = "Hello, World!"
# Accessing the first character
first_character = my_string[0]
print("First character:", first_character)   # Output: H
# Accessing the fifth character
fifth_character = my_string[4]
print("Fifth character:", fifth_character)   # Output: o
# Accessing the last character
last_character = my_string[-1]
print("Last character:", last_character)   # Output: !
# Accessing characters using slicing
substring = my_string[7:12]
print("Substring:", substring) # Output: World
```

**Note:** *It's essential to remember that when dealing with a string, it cannot be altered. This is because all string data types are immutable. Attempting to modify characters within a string will lead to a Type error.*

**When you try to change a character in a string in Python like this:**

```python
my_string = "Hello"
my_string[0] = "J"
```

**You'll encounter the following error message:**

Type Error: **'str'** object does not support item assignment.

This error happens because strings in Python are immutable, which means you can't alter characters within a string once it's created. If you want to modify a string, you should create a string with the changes you want.

For instance, to create a modified string, you can do the following:

```python
my_string = "Hello"
my_string = "J" + my_string[1:]
print(my_string) # Output: "Jello"`
```

## String Formatting

Python's string formatting operator is simplified with the **modulus (%)** operator.

```
name = "Alice"
age = 30
height = 5.8
```

```
# Using modulus % operator for string formatting
formatted_string = "Name: %s, Age: %d, Height: %.2f feet" % (name, age, height)
print(formatted_string)
```

**Output:**

```
Name: Alice, Age: 30, Height: 5.80 feet
```

**In this code example:**

- **%s** is a placeholder for a string (name in this case).

- **%d** is a placeholder for an integer (age in this case).

- **%.2f** is a placeholder for a floating-point number with two decimal places (height in this case).

The placeholder is replaced by values and provided after the % operator in the same as the placeholder appears in the string.

**String Manipulation:**

When working with strings manipulation in Python, you'll have to get involved with various operations like concatenation tasks like combining strings, cutting them into parts, adjusting their format, replacing elements, and more. Here's a quick rundown with some examples:

*1. **Concatenation**:* Combining Strings This involves merging strings by using the symbol.

```
str1 = "Hello"
str2 = "World"
result = str1 + " " + str2    # Add a space between the words
print(result)
```

**Output:**

```
Hello World
```

*2.**String Formatting**:* You can use placeholders like {} along with the format () method or f strings (to Python 3.6+).

**Example:**

```
name = "John"
age = 30
formatted_str = "My name is {}. I am {} years old.".format(name, age)
print(formatted_str)    # Result: My name is John. I am 30 years old.
# Using f-strings
formatted_str = f"My name is {name} and I am {age} years old."
print(formatted_str)    # Output: My name is John, and I am 30 years old.
```

*3. **Slicing**:* you can extract parts of a string by specifying indices.

**For instance, if you have the string:**

```
s = "Hello World"
print(s[0:5])  # Output: Hello
```

```
print(s[6:])    # Output: World
print(s[::-1])  # Output: dlroW olleH (reversed)
```

**4. Length**: When it comes to finding the length of a string, simply use the 'len()' function.

**For example, with the string:**

```
s = "Hello World,"
print(len(s))
```

**Output:**

```
12
```

**5. Replacing**: Replacing substrings within a string is easy—just use the replace method.

```
s = "Hello World"
new_s = s.replace("World", "Python")
print(new_s)
```

**Output:**

```
Hello Python
```

**6. Splitting:** To split a string into a list of substrings, use the split method.

```
s = "Hello World"
parts = s.split(" ")
print(parts)
```

**Output:**

```
['Hello', 'World']
```

**7. Join:** If you want to combine a list of strings into one string, use the join method.

**For example, if your list is:**

```
parts = ['Hello', 'World']
s = " ".join(parts)
print(s)
```

**Output:**

```
Hello World
```

**8. Case Conversion** for case conversion in a string.

```
s = "Hello World"
print(s.lower())   # Output: hello world
print(s.upper())   # Output: HELLO WORLD
print(s.title())   # Output: Hello World
```

Here is a Python code snippet that demonstrates string manipulations using Python. The code showcases how to convert a string to lowercase, uppercase, and title case. Additionally, there exist functions and methods within Python for effectively handling strings.

```python
# Sample string
text = "Hello, Python World!"
# Convert to lowercase
lowercase_text = text.lower()
print("Lowercase:", lowercase_text) #Output: hello, python world!
# Convert to uppercase
uppercase_text = text.upper()
print("Uppercase:", uppercase_text) #Output: HELLO, PYTHON WORLD!
# Convert to title case
titlecase_text = text.title()
print("Titlecase:", titlecase_text) #Output: Hello, Python World!
```

**Output:**

**Explanation:**

- **lower()**: Converts the entire string to **lowercase**.

- **upper()**: Converts the entire string to **uppercase**.

- **title()**: Converts the string to **title case**, where the first letter of each word is capitalized.

***

**Exercises**

1. Take an integer value from the user and convert it to its binary representation.

2. Create a Python program that computes the area of a rectangle. Prompt the user to input the length and width of the rectangle.

3. Write a program to convert a given temperature in Fahrenheit to Celsius.

4. Write a Python program to calculate the square root of a given floating-point number without using built-in functions.

5. Write a Python program to concatenate two strings without using the + operator.

6. Write a Python program to extract characters from index 2 to index 5 (inclusive) of a given string using slicing.

# Chapter 6: Understanding Control Structures and Functions in Python

When learning about Control Structures and functions in Python, it's important to understand the comparison operations that Python provides. These operations help in comparing values and returning either a true or false outcome based on the validity of the comparison. Special Boolean values like True and False are used in Python to help programs make decisions. Here are some fundamental comparison operators in Python:

**1. Equal** to (==): This operator verifies whether two values are equal.

```
x = 5
y = 5
print(x == y)  # Output: True
```

**2. Not equal** to (!=): This operator determines if two values are not equal.

```
x = 5
y = 3
```

```
print(x != y)  # Output: True
```

**3. Greater than (>):** This operator checks if the left operand is greater than the operand.

```
x = 5
y = 3
print(x > y)  # Output: True
```

**4. Less than (<):** This operator assesses if the left operand is lesser than the operand.

```
x = 5
y = 3
print(x < y)  # Output: False
```

**5. Greater than or equal to (>=):** Checks if the left operand is greater than or equal to the right operand.

```
x = 5
y = 5
print(x >= y)  # Output: True
```

**6. Less than or equal to (<=):** Checks if the left operand is less than or equal to the right operand.

```
x = 5
y = 3
print(x <= y)  # Output: False
```

These are the basic comparison operators in Python that can be used to compare numbers, strings, and other data types.

### Control flow statements

Control flow statements are an important aspect of Python programming. They are used to write simple code that is easy for beginners to follow. Programs usually follow a sequential structure to avoid complexity, but it can be challenging to develop programming logic linearly.

**For example:**

```
b = 3
Output: 3
```

## Conditional statements

Are a powerful programming technique that can significantly enhance the efficiency of your code. They allow you to execute only a specific part of the program based on certain conditions, thereby saving valuable time and resources. In Python, the two most commonly used conditional statements are the **"if"** and **"if-else"** statements

## Looping structures

In programming, looping structures are used to repeat lines of code or logical operations until a specific condition is met. To implement looping structures effectively, programmers must define both the starting and ending points of the loop. In Python, two commonly used looping structures are 'while' and 'for' loops, which help developers simplify their code.

## If / else Conditional Statements

Python programming involves using statements to establish the flow within a program. These control structures provide programmers with the ability to make decisions, iterate through data, and enhance code robustness. Let's take a closer look at Python's control structures and functions. Conditional statements are crucial in executing code blocks based on predefined conditions. The primary conditional statement in Python is the 'if' statement, which triggers a block of code only if a particular condition holds. Programmers can create decision-making processes by incorporating "**elif**" (else if) and 'else' statements along with multiple 'if' conditions. The 'if' statement is a tool for executing condition-based code in programming by allowing specific code blocks to run when certain conditions are met.

The basic syntax for an '**if' statement** is as follows:

## Syntax:

If condition:

    execute statement

**else:**

execute statement

**Note:** When a certain condition is met, an action is taken. If the condition is not met, a different action is executed.

**Let's observe how this code works:**

```python
# Define a variable (for example, age)
age = 18
# Check if the condition is met
if age >= 18:
print("You are eligible to vote!")
else:
print("You are not eligible to vote yet.")
```

**Here's an explanation:**

- **"if"** is used to indicate the start of a statement.

- **"condition"** refers to an expression that can result in either **True** or **False.** If it evaluates to True, the associated code block runs; otherwise, it gets skipped.

- The **colon (:)** marks the end of the if statement introduction. Indicates that a code block follows.

- **Indentation:** The code block following the if statement must be adequately indented to structure Python code. Each indentation level should contain four spaces required for Python code to run correctly. Python uses indentation to define blocks of code, unlike other languages that use curly braces **{}**.

**Why Indentation is Necessary:**

- **Defines Code Blocks**: In Python, indentation indicates which statements belong to a particular block of code, such as loops, conditionals, and function definitions.

- **Improves Readability**: It makes the code structure clear and easy to understand.

- **What Happens if Indentation is Incorrect?**

- **Indentation Error**: If indentation is missing or inconsistent, Python will raise an Indentation Error.

- **Logical Errors**: Incorrect indentation might not always cause a syntax error but can lead to unintended behavior.

**Example of Proper Indentation:**

```
def greet():
print("Hello, World!")   # This line is indented properly
greet()   # This is outside the function block, so no indentation is needed
```

**Example of Incorrect Indentation:**

```
def greet():
print("Hello, World!")   # Missing indentation
greet()
```

**Error Output:**

```
IndentationError: expected an indented block
```

**if the condition:**

# Instructions to execute if the condition is true

**else:**

# Instructions to execute if the condition is false

*Moreover, you can utilize elif (short for "else if") to incorporate additional conditions for evaluation:*

**if condition1:**

# Instructions to execute if condition1 is true

**elif condition2:**

\# Instructions to execute if condition1 is false and condition2 is true

**else:**

\# Instructions to execute if both condition1 and condition2 are false

*These are the principles of utilizing if statements in Python. They play a role in creating structures and determining actions in your programs.*

Here's a visual representation of how if, elif, and else statements work in Python using a flowchart-style diagram in text:

**(condition1?)**

   / \

  **Yes  No**

  /   \

**Execute   (condition2?)**

**Code1**    / \

    **Yes  No**

   /  \

  **Execute   Execute**

   **Code2   Code3**

**If Elif else**

In Python, **Elif is** short for **"else if."** It's used to introduce an **"if"** scenario within a sequence of statements, which is helpful when dealing with conditions that require handling.

*Here's a simple example to demonstrate how it works:*

```
x = 10
if x > 10:
print("x is greater than 10")
elif x == 10:
print("x is exactly 10")
else:
print("x is less than 10")
```

**In this scenario:**

- When if statement check if **x > 10**. If True, it executes the first block.

- If **x > 10** is **false**, it moves to **elif**, which would check if **x == 10**.

- If none of the conditions is **True, the else** block executes.

**Loops**

**Loops** are utilized to go through sets of data or execute a code block repeatedly. Python supports two loop types:

**For Loops:** The **for** loop iterates over a sequence of items (such as a list, tuple, dictionary, string, or range).

```
for i in range(5):
print(i)  # Outputs: 0 1 2 3 4
```

**Functions:**

Functions are essential for organizing your code and enhancing its reusability. In Python, you can create functions using the keyword. Here are some key aspects to note about functions:

**Defining a function**: To create a function, in Python, use the keyword followed by the function name and any parameters it requires.

You can also indicate a result to be returned by using the return keyword.

```
def greet(name):
return f"Hello, {name}!"
```

```
print(greet("Alice")) # Example function call
```

In the code snippet, we have a function called **greet** that takes a name as input and returns a greeting message using **f string** formatting.

To execute a function, after creating it, you can use it in different parts of your code to perform the operations defined within it.

```
result = greet("Alice")  # Calls the function with "Alice"
print(result) # Output: Hello, Alice!
```

Let's discuss **for loops** and their basic syntax in Python. A **for loop** is used to iterate through a sequence such as a **list, tuple, string,** or **range**. It executes a set of instructions for each element in the sequence. You can use a **for loop** or a **while loop** to check a condition and simplify this process. Here's how you can write a basic **for loop** in Python:

**for** item in **sequence**:

```
# block of code to execute for each item
```

**Let's break down each component:**

- **for:** keyword indicates the start of a for loop statement.

- **item:** This variable represents the element within the sequence. You are free to choose any name here.

- **in:** This keyword separates the name from the collection being iterated over.

- **sequence:** This refers to the set of items over which the loop iterates.

You might have a list, tuple, string, range, or any other object that can be iterated over. When using a loop, you can carry out actions on each element in the sequence.

**Here's a basic example to show how for loops operate:**

```
# Iterating over a list
fruits = ["apple", "banana", "cherry"]
for fruit in fruits:
```

```
print(fruit)
```

In this instance, the loop cycles through each element in the fruits list. For every iteration it assigns the element to the fruit. The print(fruit) command then displays each fruit.

You can also apply for loops to iterate through objects like strings:

```
# Looping through a string
word = "Python"
for letter in word:
    print(letter)
```

This will show each character of the string word on lines.

In Python, you can merge **for loops** with built-in functions like **range()**. This can help simplify and streamline your code.

```
# Using range() with a for loop
for i in range(5): # Goes through numbers 0 to 4
    print(i)
```

**This loop will display numbers from 0 to 4.**

*For looping over items well:*

When going through the items in a dictionary, here's an example:

```
# Iterating over dictionary items
person = {"name": "Alice", "age": 30, "city": "New York"}
for key, value in person.items():
    print(key, ":", value)
```

This code snippet will display each key-value pair stored in the 'person' dictionary.

In essence, using **for loops** in Python is a way to go through sequences and carry out operations on each item in those sequences. They play a role in programming tasks and are commonly used in Python scripts.

Let's talk about the **while loop** now.

**While Loop**

While the **for loop** is handy for automating tasks it does have its limitations. One such limitation is its inability to impose conditions during iteration. This is where the **while loop** shines. The while loop starts before looping begins and checks a condition every time it loops. It allows you to repeatedly run a block of code as long as a specific condition remains true. This feature makes the **while loop** a tool for developing efficient code. The **while loop**, in Python, follows a structure:

When a certain condition is met, a block of code is executed repeatedly.

**Let's break it down:**

- **While:** This term signals the start of the **while loop.**

- **Condition:** This evaluates to either **True** or **False**. The code block runs as long as this condition remains **true.**

Actions are performed inside a loop by changing the values of the condition until the loop ends. If the initial condition is False, the loop won't execute.

Here's an explanation to illustrate how **while loops** work:

```
# Counting from 1 to 5 using a while loop
count = 1
while count <= 5:
print(count)
count += 1
```

In this example, a **while loop** is used with an initial count value of **1**. The **loop** will continue to execute as long as the count is less than or equal to **5**. Inside the **loop**, the current count is printed and then incremented by **1**. This process will continue until the count reaches **5**.

It is important to ensure that the **condition** in a **while loop** will eventually become False to prevent a situation in which the **loop** runs indefinitely and crashes your program. For example, the following code results in an **infinite loop**:

```
 # Infinite loop - avoid running this code
# This loop will keep running because the condition never changes
while True:
 print("This is an infinite loop!")
```

In this scenario, the condition **True** always stays true, causing the **loop** to run endlessly.

You can also incorporate **break**. Continue statements within a **while loop** to manage how the code executes. Break is used to exit the **loop**, while continue is utilized to skip the remaining code block and proceed to the iteration of the loop.

In Python, **loops** are used to repeat a block of code until a certain condition is met. They come in handy when you are unsure about the number of times the code needs to run. However, it is important to ensure that the **condition** eventually becomes **False** to avoid **infinite loops**.

**Break & Continue**

**Loops** are a powerful tool in programming, but they can also consume memory and sometimes cause crashes. To address this issue, Python offers two programming elements: **break and continue.**

The **break statement** is utilized to stop the **loop** and move to the line following the **loop**. When the **break statement** is encountered within a loop, it signals the end of the loop and execution of statements. In Python, the break statement is employed to exit a loop prematurely.

Here's an example demonstrating its usage:

```
# Example showing how break statement works in a while loop
count = 0
while count < 5:
print("Count:", count)
if count == 3:
break # Terminate the loop when count reaches 3
count += 1
```

In this case, the **loop** will iterate from **0** to **4**. However, once the counter reaches **3**, encountering the **break statement** will cause an exit from the **loop**. Therefore, the **output** will be:

```
Count: 0
Count: 1
Count: 2
Count: 3
```

**Continue Statement**

The **continue statement** in Python is used to skip the current iteration of a **loop** and proceed to the next iteration. Here's a very basic example:

```
# Example of using continue statement in a loop
for i in range(5):
if i == 2:
continue  # Skip the iteration when i equals 2
print("i:", i)
```

In Python, the continue statement is used to bypass the iteration of a **loop** and move on to the iteration. Here's a basic example showcasing its functionality;

In this example, the **loop** runs through numbers **0** to **4**. When it reaches **2, the loop** skips that iteration and moves to the number.

**Therefore, the output will show the following:**

```
i: 0
i: 1
i: 3
i: 4
```

***

**Exercises**

1. Write a Python program using a for loop to display numbers from 1 to 10.

2. Create a Python program that counts from 1 up to a limit provided by the user. The program should ask for the limit input. Then, show each number on a line.

3. Write a code that generates a random number between 1 and 100. Prompt the user to guess the number and provide feedback if their guess is too high or too low. Repeat the prompt and feedback process until the user guesses the number correctly. To accomplish this task, use a 'while' loop in your code with the 'random' library imported and the 'random.randint(1,100)' function implemented.

4. Write a program that creates a set of numbers based on the user input value of n. Using a for loop, generate these numbers and an if statement to check if each number is even.

5. Write a program that prints the first n Fibonacci numbers. The Fibonacci sequence is a series of numbers in which each number is the sum of the two preceding ones. The first two numbers in the series are 0 and 1. To generate the numbers, use a for loop and break the loop after n numbers have been printed.

6. Write a Python function that takes two lists as input and returns True if they have the same elements in the same order; otherwise returns False.

7. Take the year as input from the user and create a program that determines if a given year is a leap year or not.

8. Write a Python program to find the sum of all prime numbers between 1 and 100 using a nested loop structure.

# CHAPTER 7: FUNCTIONS AND MODULES IN PYTHON

Python is a widely used programming language that provides various approaches to coding. One popular method is programming for straightforward projects with minimal complexity. Its versatility allows for the utilization of components in different ways. However, working with functions can be challenging as they must always be invoked within the program. Yet mastering functions and creating programs is achievable with a few examples. Originally applied in mathematics to tackle issues, functions were later adopted by programmers to recycle written code. For instance, in apps like Picsart used for photo editing, users can crop images using a tool. The developers at Picsart integrated app elements like libraries and frameworks to craft their function for this task since image cropping involves multiple intricate steps. If they wished to introduce video cropping to the app, they could. Develop a function or enhance the existing one designed for images by incorporating additional features. Typically, opting for the option proves convenient and efficient.

**Functions** can be quite challenging to create as they must interact seamlessly with all elements within the application.

**What kinds of Functions exist?**

In programming, Functions can be broadly classified into two categories. **System functions and user-defined functions**. The main Python library offers a range of system functions that developers commonly used for tasks. For example, the 'print' function is a system function that shows a specified string on the screen.

On the side, developers design user-defined functions tailored to their software needs. Users also have the option to include third-party libraries and user-defined functions in their code. Regardless of the code type utilized, the key objective for programmers when using functions is to solve problems through code.

**How does it operate?**

The concept behind employing functions in programming resembles functions. Initially the developer defines a function with logic. Assign it a name. This function can then be invoked from any part of the program using parameters. Programming elements. The developer also specifies the types of parameters users can input to prevent errors.

For a software element to perform correctly calling the function programmatically is essential.

Function invocation is commonly presented on the end using buttons, tabs, and various graphical interfaces. While it may seem simple for the user to tap on these elements behind the scenes, invoking a function involves programming it.

System functions come pre-built eliminating the need for defining them; they can only be called upon. Although developers have the option to alter system functions it is advised against due to their nature; tampering with them can potentially disrupt your code.

Python developers keen on revolutionizing software development can utilize the 'def' keyword to craft custom functions. Below is an example illustrating how function declaration works in Python for comprehension.

In essence, functions in programming aid in addressing issues with code. There exist two categories of functions: system-defined and user-defined ones; additionally, developers have the flexibility to incorporate third-party libraries well.

Python developers looking to create software can utilize the 'def' keyword to establish custom functions. Below is an example to aid in grasping function declaration in Python efficiently.

**Creating Functions**

In Python, defining a function involves using the **'def'** keyword specifying the function's name followed by parentheses where parameters can be included if needed. Following the parentheses, a **colon** signifies the initiation of the function block. The code detailing the function's operations should be within this block. The fundamental syntax is as follows:

```
def function_name(parameters): # Function definition
# body of the function
# code to execute
return value # (optional) return statement if required
```

**Here's a breakdown of each component:**

- **def function_name(parameters):** defines the function.

- The **body of the function** contains the indented lines of code that execute when the function is called.

- The **return statement** is optional. If you want to return a value from the function, you use it; otherwise, you can omit it.

```
def add_numbers(a, b):
return a + b
# Calling the function
result = add_numbers(3, 5)
print(result)  # Output will be 8
```

**In this scenario:**

- The **function** is named **add_numbers**.

- **(a, b)** represent the **parameters**.

- The operation within the function calculates the sum of **a + b**.

- When you call **add_numbers(3, 5)**, it returns **8**, which is then stored in **'result'** and printed.

**Function Parameters:**

**Function parameters** are the input you provide when defining a **function**. They enable passing data to the function, for performing tasks. These parameters are enclosed in **parentheses** after the **function name.**

Python functions have **two types** of parameters: **positional parameters** and **keyword parameters**.

1. **Positional Parameters**: These refer to the parameters passed to a **function** based on their position in the **function call**. It's important to provide arguments in the order as the parameters are defined in the function signature.

**For instance:**

```python
def greet(name, greeting):
print(f"{greeting} {name}!")
greet("Alice", "Hello") # Outputs: "Hello, Alice!"
```

In this scenario, **"Alice"** is assigned to the name parameter, and **"Hello"** is assigned to the **greeting** parameter.

2. **Keyword Parameters:** These are parameters explicitly named during a **function call**. They allow you to assign values to parameters regardless of their position.

**For example:**

```python
def greet(name, greeting):
print(f"{greeting}, {name}!")
greet(greeting="Hello", name="Bob") # Outputs: "Hello, Bob!"
```

In this example, the parameters are passed in a different order than they are defined in the function. Specifically, **"Bob"** is provided as the name parameter, and **"Hello"** is given as the greeting parameter. This raises the question of why Python produces its output in this manner, considering that the parameters are reversed.

**Example:**

```
def greet(greeting, name):
return f"{greeting}, {name}!"
# Calling the function with flipped parameters
print(greet("Bob", "Hello"))
```

Function Definition vs. Call:

- The **function** is defined as **greet(greeting, name)**, meaning:

- The **first argument** is assigned to greeting.

- The **second argument** is assigned to name.

- However, in the **function** call **greet ("Bob", "Hello")**:

- **"Bob"** is passed first, so it is assigned to **greeting.**

- **"Hello"** is passed second, so it is assigned to name.

**Output:**

```
Bob, Hello!
```

Python does not **"know"** that we meant for **"Bob"** to be the name and **"Hello"** to be the greeting—it simply assigns them in the order given.

**How to Fix It:**

To avoid this issue, we can:

1. Pass arguments in the correct order: **greet("Hello", "Bob")**

2. Use **keyword arguments** to specify which value goes where:

```
print(greet(name="Bob", greeting="Hello"))
```

This ensures **"Bob"** is assigned to name and **"Hello"** to greeting, regardless of order.

**Default parameter** values can also be set, which come into play when no value is provided for that parameter.

**For example:**

```
def greet(name, greeting="Hello"):
print(f"{greeting}, {name}!")

greet("Alice")      # "Hello, Alice!"
greet("Bob", "Hi") # "Hi, Bob!"
```

When you use the **function**, you input values for those parameters in the sequence as they are listed in the function definition.

**Here's an illustration:**

Let's say we have a **function** called **add_numbers(x, y)** that adds two numbers. We then run the **function** with **add_numbers(3,5)** where **3** is assigned to **x,** and **5** is assigned to **y** according to their positions. Therefore, **x** becomes **3**. **Y** becomes **5,** resulting in the sum of **x** and **y** being returned as **8**.

The order of **arguments** matters because it helps you clearly and intuitively provide **inputs** to a **function** by matching them with parameters based on their positions in the **function** definition.

**Keyword Arguments**

In contrast, **keyword** arguments are passed to a **function** using their parameter names. Unlike arguments that depend on the argument order, in the function call keyword arguments let you assign **values** to parameters regardless of their sequence.

**Here's a simple example:**

```
def greet(name, message):
print(f"{message}, {name}!")
greet(name="Alice", message="Hello") # Output: "Hello, Alice!"
```

In this example, name and message serve as parameters for the **greet** function. When we call the greet function, we assign values to these parameters using their names **(name="Alice" and message="Hello")**. The order in which we provide the arguments is not significant because we've explicitly specified the **parameter names.**

**Using keyword arguments provides benefits:**

**1. Clarity**: Employing keyword arguments can enhance the readability and understanding of function calls for functions with multiple parameters.

**2. Flexibility**: They enable you to assign values to the parameters you wish to specify by utilizing their names. This feature is particularly advantageous for functions with default parameter values.

**3. Error Prevention:** By utilizing keyword arguments you minimize the likelihood of mixing up argument orders in function calls, which can result in challenging to detect bugs.

**Keyword arguments** can be combined with arguments; however, positional arguments must always precede keyword arguments.

**Here's an Example:**

```
greet("Bob", message="Hi") # Output; "Hi, Bob!"
```

In this scenario, we use **"Bob"** as the name in the **function argument** and **"Hi"** as the message in the keyword argument.

In general, **keyword arguments** offer a way to define **inputs** for a function in an adaptable manner.

**Setting Default Values**

**Default values** are predetermined values assigned to parameters within a function definition. These values come into play when the function is invoked without specifying values for those parameters. This feature allows you to create functions with parameters, establishing a default behavior if the caller does not explicitly provide values for those parameters.

**Here's a simple illustration:**

```
greet("Bob", "Hi") # Output: "Hi, Bob!"
```

**In this example,** "Hi" is used as the default value **"Hello"** for the message parameter. Default values come in handy when you want to offer flexibility in your functions while still having defaults. They enhance the versatility and user-friendliness of your functions as callers can decide whether to specify parameters or stick with the defaults.

**Function Scope**

It's crucial for developers to grasp the **types of functions** and how to use them efficiently. Functions like variables have both global scopes. **Local scope variables** are created within a **function**. Are only accessible within it. Any variable that can be accessed is considered a global scope variable.

It's worth noting that a function can include both global variables. Therefore, all variables used in a function must be either local or global. The scope mechanism is primarily employed to manage memory effectively. Unused variables are typically cleared out over time to optimize program performance.

When a **function** is called, it can **recreate** the **variables**. This process still uses up runtime.

**Global scope variables** are likely to be called times once created. This makes having a **global scope** beneficial as it eliminates the need to reinitialize variables. Using scope whenever possible, regardless of the software being developed, can enhance efficiency when working on projects.

**Understanding Local and Global Scope**

In Python, variables have scopes that dictate where they can be accessed within the code. The primary scopes are global scopes.

**1. Local Scope:**

- Variables defined within a function have a local scope.

- They are only accessible within the function where they are defined.

- They cannot be accessed outside of the function.

**Example illustrating scope:**

```
def my_function():
x = 10  # The variable x has a scope inside my_function
print("Within the function x is:", x)
```

```
my_function()
# Attempting to access x outside the function will lead to an error
# print("Outside the function x is;" x) # This will raise a NameError
```

## 2. Global Scope:

Variables defined outside any function or in the scope have scope.

**Global variables** can be utilized throughout the code from any location. Within functions, these variables can be. Adjusted using the keyword. Here's an example of how global scope works;

```
x = 10 # Variable x has a global scope

def my_function():
global x #Using the global keyword to access the global variable x
print("Inside the function, x is:", x)

my_function()
print("Outside the function, x is:", x)
```

**The result will show:**

```
Inside the function, x is:  10
Outside the function, x is: 10
```

Nevertheless, it's commonly viewed as better practice to steer clear of relying on global variables since they can complicate code comprehension and upkeep. It's typically more advisable to pass **variables** as **arguments** to **functions if** they are required within those **functions**.

### A Local Scope variable can't be used in a Global Scope

When you declare **variables** inside a **function**, they have a scope, meaning they are only accessible within that function. Let me show you an example to illustrate why a variable with a local scope cannot be used in a **global scope**:

```
def my_function():
local_variable = "I am local"
my_function()
print(local_variable) # This will raise a NameError because local_variable is not
defined in the global scope
```

In this scenario, the variable `local_variable` is created within the `my_function()` function. Once the function finishes running, this `local_variable` ceases to exist. It cannot be accessed from outside the function. Therefore, if you attempt to print out `local_variable` in the scope, Python will raise a **NameError** as it does not recognize this variable.

The primary reason why **local variables** are not usable in the scope in Python is for encapsulation purposes. To prevent potential conflicts with naming. By confining scopes, Python ensures that **variables** defined in one section of code do not inadvertently impact sections. This approach improves code clarity and manageability and minimizes the risk of errors stemming from clashes or unintended alterations of names.

**Please note that all functions can access any variables within their scope without limitations; it is not restricted by their function boundaries.**

Let's take an example where a function creates and uses a variable, and then we'll try to access that variable from the scope.

```
def my_function():
local_variable = "I am local"
print("Inside the function, local_variable is:", local_variable)
my_function()
print("Outside the function, local_variable is:", local_variable) # This will raise a
NameError
```

In this scenario, the **local_variable** is declared inside the **my_function()** function. When **my_function()** is executed, it shows the value of **local_variable** within its scope. However, if you attempt to show **local_variable** in the scope, Python will throw a **NameError** because it is not defined globally.

**Explanation:**

**1. Local Scope:**

- The variable named **local_variable** exists within the boundaries of the function **my_function()**.

- It has a scope, meaning it can only be accessed within its defining function (**my_function()**).

- Once **my_function()** finishes running, the variable named **local_variable** becomes inaccessible as it goes out of scope.

**2. Global Scope:**

- An effort to access **local_variable** outside of **my_function()**.

- When a variable, like **local_variable**, is not defined in the scope in Python, a **NameError** occurs because Python cannot locate a variable by that name.

The reason why variables created within a function cannot be utilized in the global scope is to uphold encapsulation and prevent consequences. Each function should run independently with its set of variables without impacting the variables in the program. This aids in crafting manageable and less error code.

**Please note:** that when a function uses a variable, that variable is confined to that function and cannot be accessed by other functions.

**Let's demonstrate** this with an example where a local variable is declared within one function, and then an effort is made to retrieve it from another function:

```
def function1():
local_variable = "I am local to function1"
print("Inside function1, local_variable is:", local_variable)

def function2():
print("Inside function2, trying to access local_variable:", local_variable) # This will
raise a NameError
function1()
function2() # This will raise an error
```

In this example, the **local_variable** is established within **function1()**. There's an attempt to access it within **function2()**.

When you encounter a **NameError,** it's likely because the **local_variable** is not defined within the **function2()**. **Here's why**:

**1. Local Scope:**

- **local_variable** is specifically defined within the function named **function1()**.

- It has a **scope**, meaning it can only be accessed within that function **(function1())**.

**2. Accessing from Another Function:**

- In the **function2()**, there's an attempt to access **local_variable**.

- However, since **local_variable** isn't defined within the scope of function2(), Python throws a **NameError** because it can't locate a variable by that name in **function2()**.

The reason behind this restriction is to promote encapsulation and modularity. Each function should work independently with its variables without relying on or interfering with variables from functions. This practice helps in creating manageable and easily understandable code by reducing dependencies and potential conflicts between parts of the program.

**Modules:**

In programming a module consists of functions that can be utilized in any software component by importing the module and calling the functions with your parameters as inputs. Python offers enhanced module importation compared to languages like C and C++. Many developers opt to bring in modules to leverage their functions and enhance their capabilities.

One example of something Python can import that C and C++ cannot in the same straightforward way is **the random module.**

**Example in Python:**

Python has a built-in **random** module that provides easy-to-use functions for generating random numbers:

```
import random
print(random.randint(1, 10)) # Random integer between 1 and 10
print(random.random())      # Random float between 0 and 1
```

This requires **no additional setup**, making it easy for beginners.

**C and C++:**

In contrast, **C and C++** do not have built-in modules like Python. To generate random numbers, you need to use **header files** and manually seed the random number generator.

**Example in C:**

```
#include <stdio.h>
#include <stdlib.h>
#include <time.h>
int main() {
srand(time(NULL)); // Seed the random number generator
printf("%d\n", rand() % 10 + 1); // Random integer between 1 and 10
return 0;
}
```

**Example in C++:**

```
#include <iostream>
#include <cstdlib>
#include <ctime>
int main() {
srand(time(0));
std::cout << (rand() % 10 + 1) << std::endl;
return 0;
}
```

**Why Does This Matter?**

1. **Ease of Use** – Python's **import random** is beginner-friendly, while **C/C++** requires more setup.

2. **Less Boilerplate Code** – Python abstracts away complexities like **srand(time(NULL))** needed in **C/C++**.

3. **Readability** – Python's syntax is simpler and easier for beginners to understand.

This is just **one example** of why beginners find Python easier—because it has **built-in modules** for many common programming tasks that C and C++ require more effort to achieve.

In Python, a **module** refers to a file that includes Python definitions and **commands**. The module name is the filename, with the suffix **.py** appended. **Modules** can encompass functions, classes, and **variables**.

Let's craft a module called **my_module.py** that features a function for displaying a greeting:

```
my_module.py:
def greet(name):
    print(f"Hello, {name}!")
```

To utilize this **module** in another Python script you can import it using the import statement.

**For instance:**

```
main_script.py:
import my_module
my_module.greet("Alice")
```

Upon executing **main_script.py**, it imports **my_module** and invokes the greet function within it to print **"Hello, Alice!"**.

There are **methods** for importing from a **module**:

1. Importing the **module** and accessing its contents using dot notation:

```
import my_module
my_module.greet("Alice")
```

2. Importing functions or variables from a module:

```
from my_module import greet
greet("Alice")
```

3. Importing all functions and variables from a module (not recommended due to name clashes):

```
from my_module import *
greet("Alice")
```

**To make a module:**

1. Compose the Python code with the functions, classes, or variables you wish to include in the module within a.py file.

2. Ensure that the file name is a Python identifier (like my_module.py).

3. Store the file in a location where Python can access it (in the folder as your script or in a directory listed in the Python path).

Once you've created your module, you can. Utilize its contents in Python scripts as demonstrated in the examples provided.

**Modules and Built-In Functions**

Modules and built-in functions are essential for code organization, reusability, and efficiency. Let's take a closer look at each concept:

**Modules:**

A module in Python is essentially a file that contains Python definitions and statements. The name of the file becomes the module name, with.py added as a suffix. Modules can define functions, classes, and variables that are reusable by importing them into Python scripts.

**How to Formulate a Module:**

You can establish a module by scripting Python code within a **.py** file. For instance, if you create a file named **mymodule.py** comprising this code snippet:

```python
# mymodule.py
def greet(name):
print("Hello, " + name)
```

To add **two numbers**, in Python, you can define a function like this:

```python
def add(a,b):
return a + b
```

Here's how you can utilize a module:

Once you've created a module you can incorporate it into Python scripts by using the import statement.

```python
import mymodule
mymodule.greet("Alice")
result = mymodule.add(3, 5)
print(result) # Output: 8
```

You also have the option to import functions or variables from a module directly:

```
from mymodule import greet
greet("Bob")
```

## Built-in Functions:

Python includes a range of built-in functions that are readily accessible without requiring any module imports. These functions are part of the Python library. Offer fundamental operations and features.

Some used built-in functions are;

- **print():** Used for displaying objects on the console.

- **len():** Returns the length of an object such as a string, list or tuple.

- **range():** Creates a sequence of numbers.

- **input():** Captures user input.

- **type():** Indicates the type of an object.

- **sum():** Computes the sum of elements in an iterable.

## Here's an example usage:

```
print("Hello, world!")
print(len("Python"))
for i in range(5):
print(i)
name = input("Enter your name: ")
print("Hello,", name)
print(type(42))
print(sum([1, 2, 3, 4, 5]))  # Output: 15
```

These are a few examples of built-in functions; Python offers many more. You can refer to the Python documentation.

For a list of defined operations. Both libraries and pre-defined operations are essential for writing Python code effectively promoting code reusability organization and readability.

**String Functions**

Python provides a variety of string functions that allow you to manipulate and work with strings effectively. Here are some basic string functions:

**1. len():** Provides the length of the string.

```
my_string = "Hello, world!"
print(len(my_string)) # Output: 13
```

**2. capitalize():** Returns a version of the string with the letter capitalized.

```
my_string = "hello, world!"
print(my_string.capitalize())  # Output: "Hello, world!"
```

**3. upper():** Returns a version of the string, with all letters converted to uppercase.

```
my_text = "hello, world!"
print(my_text.upper()) # Output: "HELLO, WORLD!"
```

**4.lower():** Returns a copy of the string with all characters and converts the string to lowercase.

```
my_string = "HELLO, WORLD!"
print(my_string.lower()) # Output: "hello, world!"
```

**5.strip():** Returns a copy of the string and removes leading and trailing whitespace.

```
my_string = " hello, world! "
print(my_string.strip()) # Output: "hello, world!"
```

**6.replace():** Returns a copy of the string with all occurrences of a substring replaced with another substring.

```
my_string = "hello, world!"
print(my_string.replace("hello", "hi")) # Output: "hi, world!"
```

To define a string variable `my_string` as **"hello, world!"**. Replace all instances of **"hello"** with **"hi"** using the `replace()` function. The output will be **"hi, world!"**.

**7.split():** Split string into a list of substrings based on delimiter.

```
my_string = "apple,banana,orange"
fruits = my_string.split(",")
print(fruits) # Output: ['apple', 'banana', 'orange']
```

**8.join():** Join elements of a list into a single string using the string as a separator.

```
fruits = ['apple', 'banana', 'orange']
my_string = ",".join(fruits)
print(my_string) # Output: "apple,banana,orange"
```

**9.find():** Returns the lowest index of the substring if found in the string, **-1** otherwise.

```
my_string = "hello, world!"
print(my_string.find("world")) # Output: 7
```

**10.count():** Returns the number of occurrences of a substring in the string.

```
my_string = "hello, hello, hello, world!"
print(my_string.count("hello")) # Output: 3
```

Python's standard library provides a range of built-in functions for manipulating strings. Among these are **find()** for finding the index of a substring.

***

**Exercises**

1.Write a Python function to calculate the factorial of a given number using recursion function.

2. Create a Python module named **math_operations** that contains functions for addition, subtraction, multiplication, and division. Demonstrate how to import and use these functions in another Python script.

3. Write a Python function **count_vowels** that takes a string as input and returns the count of vowels (a, e, i, o, u) in the string. Test your function with various input strings.

4. Implement a Python function **reverse_list** that takes a list as input and returns a new list with the elements reversed.

5. Define a Python function **is_palindrome** that takes a string as input and returns True if the string is a palindrome (reads the same forwards and backward); otherwise returns False.

6. Write a Python module named geometry that contains functions to calculate the area and perimeter of a rectangle. Create another Python script to import this module and use these functions to calculate the area and perimeter of a rectangle with given dimensions.

# CHAPTER 8: OBJECT ORIENTED PROGRAMMING (OOP)

Exploring Object-Oriented Programming in Python Object-oriented programming **(OOP)** is a coding approach that emphasizes structuring code into objects that contain both data and actions. In Python, every element is viewed as an object. **OOP** is widely utilized for crafting organized code. Classes play a role in the realm of object-oriented programming within Python. A **class** serves as a template for generating objects specifying their characteristics (attributes) and behaviors **(methods)**. To define a class in Python, you utilize the class keyword followed by the designated class name.

**An illustration of defining a class in Python:**

```
class Car:
def__init__(self, make, model, year):
self.make = make
self.model = model
self.year = year
def display_info(self):
print(f"{self.year} {self.make} {self.model}")"
```

**In this example**, we define a class named **Car** that contains various attributes like make, model, and year. We also define a method named **display_info** to print out the car's details. Furthermore, we also define a special method __init__ which is used to create an object (also known as an instance) of the class.

To create an **instance** of a **class,** you can call the class like a **function**. Here's an example of creating a Car class:

```
my_car = Car("Toyota", "Corolla", 2020)
my_car.display_info()  # Output: 2020 Toyota Corolla
```

**Inheritance** is a fundamental concept in object-oriented programming (OOP). It allows you to create new classes based on existing ones, called the superclass. The subclass inherits all the attributes and methods of its superclass while also having the capability to override or extend them. Here's an example of how inheritance works in Python:

```
class ElectricCar(Car):
def __init__(self, make, model, year, battery_type):
super().__init__(make, model, year)
self.battery_type = battery_type
def display_info(self):
super().display_info()
print(f"Battery Type: {self.battery_type}")
```

**In this example,** we create a subclass named **ElectricCar** that inherits from the Car class. The **ElectricCar** class adds a new attribute called **battery_type** and modifies the **display_info** method to include information about the battery type.

**Python's object-oriented programming** provides a way of structuring and organizing code, thereby simplifying the management of complex systems and promoting code reusability. By understanding and applying concepts such as **classes, objects, inheritance,** and other **OOP** principles in Python programming, you can develop flexible and adaptable code.

**Python supports inheritance**, which allows a **class** to inherit attributes and methods from its parent classes. In certain scenarios, it can be useful for a class to combine functions from multiple sources. However, using inheritance can make class structures more complex and potentially create issues such as the **diamond problem**.

Python provides **two features**, **method overloading** and **method overriding**, to increase class flexibility and adaptability. **Method overloading** involves defining **methods** with the same name but different **parameters**, although Python doesn't support **true method** overloading as some other programming languages do. On the other hand, method overriding involves redefining a method in a **subclass** to provide an implementation while keeping the method signature as the superclass.

The diamond problem occurs in multiple inheritance when a class inherits from two classes that both inherit from a common base class. This can create ambiguity about which method should be called. Python uses the **Method Resolution Order (MRO)** and the **C3 linearization algorithm** to handle this issue. Here's a Python code snippet demonstrating the **diamond problem**:

```python
class A:
def show(self):
print("Class A")
class B(A):
def show(self):
print("Class B")
class C(A):
def show(self):
print("Class C")
class D(B, C):  # Multiple inheritance
pass
d = D()
d.show()  # Resolves using MRO
print(D.__mro__)  # Prints the method resolution order
```

**Explanation:**

- **A** is the base class.

- **B** and **C** both inherit from **A**, creating two branches.

- **D** inherits from both **B** and **C**, forming a diamond shape.

- When calling **d.show()**, Python resolves the method using **MRO**, meaning it will first look at **B**, then **C**, then **A**.

**Output:**

```
Class B
(<class '__main__.D'>, <class '__main__.B'>, <class '__main__.C'>, <class '__main__.A'>, <class 'object'>)
```

Python resolves method calls using **depth-first, left-to-right order**, avoiding duplication using the **C3 linearization algorithm**.

Python supports **encapsulation** by grouping data **(attributes)** and **methods** that manipulate the **data** within a **class**. It provides access control to that data by using public, protected, and private access modifiers. Developers can indicate the intended accessibility level of class elements by following naming conventions, such as using an **underscore** for attributes (protected) or a **double underscore** for attributes.

In Python, underscores are used to indicate the intended visibility of attributes and methods. Here's how they work:

**Single Underscore (_)** – Convention for "Protected" Attributes.

- A single underscore before a variable **(_attribute)** is a convention to indicate that it is intended to be **protected**. This means it is meant for internal use but **can still be accessed** from outside the class if necessary.

**Example:**

```
class Example:
def __init__(self):
self._protected_var = "This is a protected variable"
obj = Example()
print(obj._protected_var) # Can still be accessed, but should be avoided.
```

**Double Underscore (__)** – Name Mangling for "Private" Attributes.

- A double underscore before a variable **(__attribute)** triggers **name mangling**, meaning the attribute's name gets internally changed to **_ClassName__**at-tribute.

- This makes it more difficult (but not impossible) to access it from outside the class.

**Example:**

```
class Example:
def __init__(self):
self.__private_var = "This is a private variable"

obj = Example()
# print(obj.__private_var)  # This will raise an AttributeError
# But it can still be accessed using name mangling:
print(obj._Example__private_var)  # Accessing the mangled name
```

**When to Use Which?**

- Use **_attribute** when you want to indicate that an attribute is **not meant to be accessed directly** but still **accessible if necessary**.

- Use **__attribute** if you want to make it harder for subclasses or external code to access it, mostly to avoid accidental overrides.

**Polymorphism** is a feature in Python that allows **objects** from **various classes** to be treated as objects of a shared **superclass**. This feature enables dynamic coding, where the same method or operation can exhibit different behaviors according to the type of object it is working with. **Polymorphism**, along with **Object-Oriented Programming (OOP)** concepts like abstraction and inheritance, empowers programmers to develop scalable software systems with Python.

\*\*\*

**Exercises**

1.How does Python support multiple inheritance? Provide an example.

2. Give an example of using encapsulation to protect class attributes from direct access.

3. Give an example of using inheritance and encapsulation together to create a specialized class.

4. Explain how polymorphism allows for greater flexibility in Python programming.

5. Create a Python class called **BankAccount** that encapsulates the balance of an account. Implement methods **deposit()** and **withdraw()** to modify the balance, ensuring that the balance cannot be accessed directly from outside the class.

6. Define a Python class called Shape with a method **area()** that returns 0. Then create two subclasses, Rectangle and Circle, that inherit from Shape and override the **area()** method to calculate the area of a rectangle and a circle, respectively.

# CHAPTER 9: FILE HANDLING AND INPUT/OUTPUT OPERATIONS

Working with Files and Input/Output Tasks Managing files in Python is made easy with built-in functions and methods, allowing for smooth reading and writing of data.

It's crucial to understand the different modes used when opening files in Python using the **open()** function. The function requires the file path and mode as parameters. The available modes for opening files are:

- The('**r**') mode allows you to read data from a file.

- The ('**w**') mode lets you write data to a file, replacing any existing content.

- The ('**a**') mode allows you to append data at the end of the file.

- The('**b**') mode is useful for text files to ensure proper file content handling.

**For example:**

```
file = open("example.txt", "r")
```

After opening a file, you can perform various operations on it. To read the file contents, you can use the **read()** method to read the entire file at once or the **readline()** method to read the file line by line. If you want to write to a file, you can use the **write()** method, while the **writelines()** method allows you to write multiple lines. In addition, if you need to process large files, you can use a **for loop** to loop through the lines of the file. This approach is quite handy when you want to perform certain operations on each line of the file.

```
# Open the file in read + write mode
with open("example.txt", "r+") as file:
  # Read from the file
content = file.read()
print("File content:", content)

  # Write to the file (adds content starting from the current cursor position)
file.write("\nHello, world!")  # Appends "Hello, world!" at the end
```

When you **write data** to a file using the **write()** method, it appends the content to the file. However, it is important to keep in mind that after performing any **write** or **read** operations on the file, you must **close the file** using the **close()** method. This is crucial to release system resources and to prevent any data loss or corruption. Alternatively, you can use the with statement, which automatically closes the file once the code block finishes executing. This makes the code more **readable** and **safe**.

```
with open("example.txt", "r") as file:
content = file.read()
print(content)
```

Python offers a variety of functionalities for file handling that can help developers manage files more effectively. One of these functionalities is the **seek()** method, which enables you to adjust the file pointer to a specific location in the file. Additionally, you can validate the existence of a file on your system by using the **exists()** function from the **os.path** module. To manage errors that may occur during file operations, you can use **try-except blocks.** You can also regulate file permissions using the **chmod()** function, which determines who can read or write to the file.

It's important to handle any errors that may arise during file operations and provide informative messages. Python provides several packages that make file management more convenient and robust. The OS package can be used for tasks related to the operating system, while the **shutil package** is designed for file tasks such as copying and moving files. You can also use the **pathlib package** to organize file paths in an object-oriented manner. By using these packages, developers can confidently perform any file operation tasks.

Python is not limited to working with just text files. It also supports other data formats, such as **CSV (Comma Separated Values)** and **JSON (JavaScript Object Notation)**. When it comes to handling structured data, the **CSV** module provides features for both reading and writing **CSV files,** making it a popular choice. On the other hand, **Python's JSON** module simplifies the process of serializing and deserializing **JSON data,** which makes it easier to share data across different systems.

It's important for developers to have a thorough understanding of **file management** and **input/output procedures** in Python because it helps in creating **scalable applications** that can efficiently interact with external data sources. By mastering the complexities of file management and following recommended practices, developers can optimize their code for better performance and easier maintenance.

# CHAPTER 10: ADVANCED PYTHON CONCEPTS AND BEST PRACTICES

In this section, we will discuss some important Python concepts and best practices that will help you improve your Python programming skills. By gaining a thorough understanding of these topics, you will be able to write more efficient, maintainable, and standardized code. Now, let's delve into each concept:

Generators are a special feature in Python that enables you to create iterators without the need to define a class with **__iter__ and __next__ methods**. When you call a generator function, it returns a **generator object** that you can iterate over using a for loop. Instead of using the return statement, you can use the yield keyword to pause the function's execution and return a value. This results in a more efficient way of generating sequences of values, as it does not require storing them in memory. This is especially helpful when working with large datasets or infinite sequences. Once you understand how generators and iterators work, you can write code that is more concise and memory-efficient.

**Iterators** are objects that allow you to access elements in a sequence one by one by implementing the iterator protocol. You can create your own custom iterators by defining **an __iter__ method** that returns self and a **__next__ method** that raises **StopIteration**

when there are no more elements left to iterate over. By understanding how generators and iterators work, you can write more concise and memory-efficient code.

**Decorators:**

In Python, **decorators** are functions that can modify the behavior of other functions. They allow you to add new functionality to existing functions without changing their source code. **Decorators** are often used for tasks such as logging, **debugging**, timing, caching, and authentication. You can easily apply a decorator to a function by using the @decorator_name syntax.

**Context Managers:**

In Python, a **context manager** is a useful tool that allows you to manage resources by defining actions to be performed before and after a block of code is executed. This is accomplished using the __enter__ and __exit__ methods, which are used to define the setup and teardown actions. Context managers are often used to handle file operations, database connections, and thread synchronization. By creating your own custom context managers using the with statement, you can ensure that resources are properly managed and cleaned up in your code.

**Custom Exceptions:**

When programming in Python, you can create custom exceptions by defining new exception classes inherited from the **built-in Exception class**. The main benefit of using custom exceptions is that they allow you to handle different types of errors in a more specific way. By raising custom exceptions when unexpected situations occur, you can provide clear and informative error messages to help users understand what went wrong. This approach can help you manage errors effectively and improve the overall resilience of your code. It's important to follow best practices when developing Python code to ensure that it's clean, readable, and maintainable.

**Some key best practices include:**

Clear variable names are not only a good practice but a necessity when programming in Python. Using names that indicate their purpose makes your code self-explanatory, reducing the time and effort needed for understanding and debugging. This practice is crucial for efficient and effective Python programming. By following these practices, you can develop high-quality Python code that's easy to understand, test, and manage. Furthermore, adhering to best practices promotes effective collaboration among developers and contributes to a structured and organized codebase. By mastering these advanced concepts and best practices in Python, you can enhance your programming skills and become a more proficient and efficient developer. Embrace these principles to write elegant, reliable, and professional code that showcases your Python programming expertise.

# CHAPTER 11: APPLYING PYTHON IN REAL-WORLD PROJECTS

In this section, we will delve into the practical applications of Python in real-world projects, showcasing its significance in various sectors and industries. Python's versatility and user-friendly syntax have established it as a preferred programming language for developers searching for solutions in diverse domains.

**Python frameworks** like **Django and Flask** have revolutionized the way dynamic and interactive websites are developed in web development. Django provides a framework that simplifies complex web development tasks, while Flask offers a more adaptable option for creating scalable web applications. The extensive use of these frameworks can be seen in numerous websites and applications that are powered by Python, including popular platforms such as Instagram, Pinterest, and Spotify.

Python boasts a remarkable strength in data analysis and visualization that is evident through its libraries such as NumPy, Pandas, and Matplotlib. These tools, with NumPy's potent array processing, Pandas' data manipulation capabilities, and Matplotlib's advanced plotting features, empower data scientists and analysts to not only extract insights from datasets but also to present them in a visually engaging manner, thereby enhancing the value and impact of their work. The use of Python, in combination with platforms such as Jupyter Notebook, improves the process of exploring and analyzing data. This feature has made Python a popular choice in various fields, such as finance, marketing, and scientific research.

Python has a vast range of tools and libraries that are incredibly useful for machine learning and artificial intelligence applications. These tools, such as Scikit Learn, TensorFlow, and Keras, make it possible to use advanced algorithms to train and deploy models easily. Python is a highly adaptable and compatible language, which makes it an excellent option for developing AI solutions. Many different industries, including healthcare, autonomous vehicles, and e-commerce, have adopted Python for various tasks, such as natural language processing, image recognition, and recommendation systems. These industries use Python to drive innovation and enhance user experiences.

**Python's automation capabilities** have transformed how businesses streamline processes to improve their efficiency. With the ability to write scripts for automating tasks or complex workflows orchestrating entire systems, Python stands out as an unparalleled automation tool. Industries such as finance and healthcare have taken advantage of Python's automation features to optimize operations and promote innovation. The growing adoption of DevOps methodologies has also emphasized Python's role in automating infrastructure provisioning, deployment processes, and monitoring activities to ensure smooth operations in modern cloud environments.

**Python's flexibility** and strong community support make it a valuable tool for developers looking to address real-world problems in different fields. By leveraging Python's capabilities and utilizing its range of libraries and vibrant community, developers can explore new opportunities and create meaningful solutions that drive progress in today's dynamic tech world.

# CONCLUSION

If you enjoy reading this book, I would greatly appreciate it if you could write a positive review on Amazon. Furthermore, I have developed numerous projects over the years, using the programming knowledge I gained from coding. I own a staffing agency and a tech company called Appacore, which specializes in developing apps, websites, software, and marketing strategies and recruiting staffing agencies. We have been in business for over four years. Additionally, I have my own social media platform called Fanisee, a space rental app called Bizitme, and I created my own cryptocurrency called Apiece API. It is a unity token that syncs to all my platforms. I have attached QR Codes for all my projects, which you can scan if you are interested. In the meantime, I will be teaching this book as an online course on Udemy. I will also be adding updates to the course based on what I've learned from writing this book, as well as some additional information that I wish I had included in the book. My goal in writing this book was to make it as simple as possible and filled with useful information for beginners or anyone interested in learning how to write code in the Python programming language. I have successfully built numerous projects and worked for top tech companies. I also hold a degree in computer science with certificates in Python, ISTQB, and AWS. I wrote this book in a simple manner so that anyone could start learning how to write Python code, and I included exercises to help you get started. I have spent a lot of time and effort creating this course and book. I created it with all my heart, especially for people like you who are eager to learn. Lastly,

I would like to mention that I have a learning disability. I am dyslexic. Therefore, if you come across any typos or errors in the code within this book, please do not hesitate to report it to my email at info@appacore.com. I am always learning and would love to learn from my mistakes.

Url:

Appacore:

www.appacore.com

Bizitme Rental space app:

https://apps.apple.com/us/app/bizitme-coworking-spaces/id1583198715

https://play.google.com/store/apps/details?id=com.appacore.bizitme&hl=pt&gl=US

Social media Platforms:

www.fanisee.com

Apiece token (API):

https://dexscreener.com/ethereum/0x3fc1d4b9aa01b80e9347adf8b27ee224f214fd82

# HOW CAN YOU HELP ME AS AN AUTHOR?

As someone who writes, I know how valuable it is to hear from readers. Your feedback can really make a difference in getting my work to more people. Writing this book has been tough yet fulfilling. Despite facing writer's block at times and trying to present information, I've pushed through. Opted to be an independent author for full control over the content. If you've already read my book and found it helpful, I'd appreciate it if you could spare a few minutes to leave a review on Amazon. Your review could help inform others about the book's merits and allow more folks to benefit from it. Remember to scan the QR code at the start of the book for access to Python interview questions, answers and exercise solutions. This will aid in enhancing your Python skills and boosting your confidence. To share your review on Amazon, scan the QR code provided below. If you're willing, consider sharing a video expressing your thoughts on the book or even a review with some images of the book. Your contribution would mean a lot to me. Feel free if you choose otherwise. I'm thrilled to have you come along with me on this journey toward mastering Python.

# ACKNOWLEDGMENTS

I would like to begin by expressing my gratitude to my older brother, Maly Mam, for his invaluable contributions to this book. He helped me come up with the title and idea for this book, designed the cover, and did a lot of the editing. Without him, this book would not have been possible.

I dedicate this book to my father, Saran Mam, who is no longer with us. Dad, your unwavering support and belief in me inspired me to write this book. I hope that it does justice to the legacy you left behind.

I would also like to acknowledge Lakmal @logodesign36h from Fiverr for designing the Python logo images.